Water Resources: A New Era for Coordination

WILLIAM WHIPPLE, JR.

1801 Alexander Bell Drive
Reston, Virginia 20191-4400

Abstract: *Water Resources: A New Era for Coordination* is a book that relates the current problems of water resources developments. Geared toward those professionals with basic engineering knowledge, it covers topics in the field, such as the environment, population and development, water supply, floods and flood control, navigation, hydroelectric power, irrigation, multiple (comprehensive) purpose planning system, water quality and pollution control, runoff control/storm water management and standards, coordination under present institutions, and coordination through federal/state planning. It stresses the need for coordination between current environmental regulations and water resources planning, though it also covers much of the history of water resources planning. This book precedes a national conference on this most important issue that will take place in Chicago in June, 1998.

Library of Congress Cataloging-in-Publication Data

Whipple, William, 1909-
    Water resources: a new era for coordination/William Whipple, Jr.
      p.   cm.
    Includes bibliographical references and index.
    ISBN 0-7844-0328-7
    1. Water resources development. 2. Water-supply—Planning.
3. Water resources development—Law and legislation. I. Title.
  TC405.W58 1998
  333.91'15—dc21   98-2737
                                      CIP

Preface

The impetus to write this book came largely from my three years of experience as consultant for the Corps of Engineers, attempting to find a solution for the problems of the Chattahoochee, Apalachicola, and related rivers in the Southeast, and working on problems of drought management in other parts of the United States. It became apparent that water resource planning activities today are basically different from what they were in past years, when I worked on the Missouri River, the Columbia River, and the Arkansas River. All of these rivers had very difficult problems to solve, but the process of getting the municipalities, states, and other agencies to agree to a solution seemed much simpler. Planners could work on the problems directly and openly with reasonable expectations of agreement. Today, various agencies and groups have agendas of their own, and they hardly seem to speak the same language. It is as though we were engaged in some great game, with no fixed rules and no referee, with various groups each seeking its own advantage. The main effort of planners has to be negotiation and public relations rather than getting on with the job of finding a solution to the problem. I came to feel that it is urgent that some better means of coordination be devised.

Gradually, after discussions, a group of prominent water resource planners came to agree with this viewpoint. The ASCE became interested, and government agencies agreed to cosponsor a national conference on the coordination between environmental regulations and water resources planning to be held in June 1998 in Chicago. I will be heavily involved in the conduct of this conference.

Meanwhile, the ASCE requested I write a book about the new era in water resources planning and management. I wish to express deep appreciation for the valuable assistance of friends and associates in preparing this book. Neil Grigg of Colorado State University reviewed preliminary drafts of the entire text and gave helpful suggestions. Harold Woods of New Jersey American Water Company gave valuable information on the new processes of water treatment by ozonation. Thomas Cawley, former president of Elizabethtown Water Company, helpfully reviewed the chapter on water supply. Richard Palmer of the University of Washington provided details on the modern use of computers in complex interagency planning. Many others gave assistance by their influence through past associations and writings and by preparing materials for the conference. Helaine Randerson did a superb job of manuscript preparation and also drew the illustrations.

However, for any errors, undiplomatic bluntness, and the novel suggestions about federal-state planning, I am wholly responsible.

William Whipple, Jr.

Contents

Preface iii

Chapter 1   Introduction 1

Concept of this Book 1

Classification of Water Uses and Effects 2

Objectives of Water Resources Development 3

Government Agencies and Their Functions 6

Chapter 2   Environment 11

Introduction 11

Wetlands Control and Stream Restoration 15

Parks and Constructed Environments 20

Water Quality 21

Environmental Effects Downstream of Dams 21

Endangered Species 22

Relative and Absolute Values 23

Chapter 3   Population and Development 27

Introduction 27

Birth Rate, Immigration, and the World Picture 27

Urbanization, Industry, and Deterioration
of Large Cities 28

Encroachment upon Open Space and Agriculture ... 28

Nonpoint-Source Pollution 29

Conclusion 29

Chapter 4   Water Supply 31

Introduction 31

Value of Benefits 32

Rivers, Lakes, and Reservoirs 32

Groundwater and Its Conjunctive Use 34

Treatment, Including Ozonation 35

Reprocessing Wastewater 37

Desalination 37

Water Supply Problems 39

Water Supply Critical Areas 40

Water Supply Drought Management 45

Chapter 5   Floods and Flood Control 51

Major Floods of the Past 51

Floodplain Management 53

Special Districts 57

Benefit Evaluation 57

Flood Control Storage 58

Runoff Pollution 59

Environmental Use of Floodplain Zoning 60

Summary 61

Chapter 6   Navigation, Hydroelectric Power, and Irrigation 63

Navigation, General 63

Current Navigation Programs 64

Benefit Values of Navigation 64

Hydroelectric Power, General 64

Hydroelectric Power Operations 65

Controversies Related to Hydroelectric Power 65

Benefit Values of Hydroelectric Power 67

Western Irrigation and Subsidies 68

Irrigation Elsewhere 69

Pollution and Groundwater Depletion
from Irrigation 70

Valuation of Irrigation Benefits 70

Chapter 7   Multiple-Purpose (Comprehensive)
Planning System 73

Introduction 73

The Objectives and Concept 75

Potential Projects and Screening Alternatives 75

Selection Process 76

Computers and Modeling 76

Present Situation 77

Chapter 8   Water Quality and Pollution Control 79

Introduction 79

Nonpoint Sources 79

Water Quality Criteria and Standards 81

Conclusion 82

Chapter 9   Runoff Pollution Control/Stormwater Management . 85

Introduction 85

Control of Runoff Pollution 85

When to Control 87

Nonpoint-Source Pollutant Removal Methods 87

Design of Dual-Purpose Detention Basins 89

Regional Detention Basin Systems 92

Federal Requirements 92

State Programs 93

Chapter 10   Coordination Under Present Institutions 97

Introduction 97

Basic Policies for Coordination 98

Controversial Aspects 99

Government Attempts at Coordination 99

Cal-Fed Bay Delta Program 100

The Alabama-Coosa-Tallapoosa and
Apalachicola-Chattahoochee-Flint (ACT-ACF) 100

Platte River 102

Water Pollution Control 102

Conclusion 102

Chapter 11   Coordination Through Fed-State Planning 105

Introduction 105

Computers in Fed-State Studies 105

The Fed-State Study 107

Consequences If No Major Action Taken 109

Glossary   111

Index   117

Chapter 1

INTRODUCTION

Concept of This Book

Water resources development went through what might be called the pioneering era for about 25 years after World War II. It was a time when the federal agencies enthusiastically planned and built great systems of projects (such as the Columbia, Colorado, and Missouri) patterned after the Tennessee Valley Authority (TVA). It was an era when then-Senator Lyndon Johnson could remark almost casually that he arranged every year for a new Corps of Engineers reservoir project to be built for Texas, and every other year a Bureau of Reclamation project. This was not just talk; the number of projects built met the schedule. States and the public welcomed the new projects, especially as the federal government bore most of the costs.

However, since 1970, the picture has entirely changed. Although needs for water supply and for flood control continue to grow, the majority of U.S. rivers suitable for water resource development have structures already built, and most other valleys are so occupied by communities and industries that reservoir construction would be impractical. Moreover, the growth of the environmental movement has brought other considerations into attention. Pollution control programs for municipalities and industries are unable to prevent increasing pollution of streams from nonpoint sources. Depletion and pollution of groundwater have become critical issues. There is increased competition between uses of water, uses of land, and the need for environmental protection. There are complex linkages between human and ecological systems. The states have major responsibilities in some of these aspects.

Today's problems are generally not solvable by the building of new large dams and comparisons of benefit/cost ratios. Increasingly, the federal construction agencies find their hands tied by problems needing solution by other federal agencies, the states, municipalities, and sometimes private concerns. Experienced federal planners may have nostalgia for the good old days in the United States, when nothing but hard work and accurate thinking were enough to lead to a solution, but those days are gone forever. The water problems are still urgent, but more sophisticated approaches with a much broader viewpoint are now necessary.

This is a book for professionals who are concerned with the new basin-wide, multi-agency problems that are now arising. The reader is presumed to have a basic knowledge of hydraulics, hydrology, laws of probability, and other basic engineering disciplines. Moreover, there are a number of specialized fields whose technology is not covered, since such projects have been widely used for years, and the agencies concerned have staffs fully competent to plan them. This includes such matters as design of dams, locks, hydroelectric power plants, and water supply distribution systems. All these subjects need to be covered in new-era planning, but only aspects relevant to coordination with other programs need to be included in this book. Books already exist covering basic technology of water resources management (e.g., Viessman and Welty, 1985).

Classification of Water Uses and Effects

Water affects mankind in many ways. Some uses of water are consumptive, such as water supply, irrigation, and many manufacturing activities. The term *water supply* primarily means domestic consumption, but it is often used to include industrial use and always includes watering of home lawns and gardens. Other uses are non-consumptive, such as navigation, hydroelectric power, fishing, cooling water for thermal power and manufacturing, and water-based recreation. Other effects of water are not uses at all, such as floods and harmful effects of pollution.

Water activities usually included in multiple-purpose or comprehensive basin-wide development include water supply, flood control, navigation, hydroelectric power, and irrigation. Studies of such activities are usually conducted by the construction agencies. Other major activities that are related in effect but managed through entirely different channels are pollution control, soil erosion and erosion prevention, coastal protection, management of fish and wildlife, wetlands management, and local runoff control. The increase in relevance of this second class of activities related to water, and the multiplicity of agencies concerned with them, have rendered planning and management of water resources much more complex, as detailed in section D.

Objectives of Water Resources Development

General objectives of water resources development were first implied by the Flood Control Act of 1936, which stated that the benefits (B), to whomsoever they may accrue, should exceed the costs (C), or more simply put in quantitative terms, $B/C > 1$. Subsequent federal guidance has amplified this criterion, with some discrepancies and omissions, which are explained in detail in the chapters that follow. This federal guidance is generally sound but leaves some important matters unstated. The following table lists the objectives that should be maximized and major problems to be avoided in water resources development. The order given does not represent priority.

Objectives of Water Resource Planning

• Maximize tangible benefits less costs

• Saving of lives

• Environmental effects (non-human)

• Liberty to live without disruption

• Welfare effects on human life, including traditional, historical, religious, and archeological aspects

The first objective is to maximize the tangible direct benefits from the project, as compared to the tangible costs of building and maintaining it. Since most of the costs are incurred at the time of construction and the benefits occur over a period of many years after the project is finished, a discount rate (usually the rate of return of long-term federal bonds) is used to compare dollar value of costs at time X to benefits at time Y. However, this comparison is valid *only* to the extent that the value of money (the general price level) has not changed during the period in question. Correction of the estimate of flood damages controlled or other project benefits at a later date corresponding to price level changes is often overlooked, in effect increasing the benefit/cost ratio unduly. However, the B/C ratio determined at the time the decision is made to build a project avoids this fault, if its projected benefits are based upon an assumption of unchanged price levels for the future.

The second important factor is life itself. This point arises mostly in flood control projects, which may save lives. Sometimes attempts are made to put a money value on human life. This is unacceptable in principle, but rough approximate evaluations are sometimes made.

The third group of objectives and problems is the broad category of environmental effects. These effects may be benefits but more often are projected adverse effects of the proposed (or actual) project. These include effects on fish and wildlife generally, endangered species of plants, fish, birds and animals, and also the value of the area for public recreation. The evaluation of environmental effects is an enormously complex matter, covered in detail in Chapter 2. An illustration is the Cherry Creek Dam, near Denver. It was originally planned for flood control only. However, local interests soon recognized its potential recreational value and solved the water rights question. It now provides a beautiful recreational lake in that arid countryside.

The fourth important factor is the liberty of people to lead the lives they wish without undue disturbance. In many parts of the country, families have had their lives disrupted by prolonged drought or by floods, which destroy their way of life and may force them to leave and live elsewhere. More rarely, a similar disruption may occur through erosion. In many cases, water resources developments can provide safety against such eventualities. There are many less disastrous but still serious events that flood control and assured supplies of water can guard against. Prudent families are usually prepared to face unexpected losses of moderate amount, but to protect against major losses (such as destruction of an automobile or death of a wage earner) prudent families buy insurance, often paying a great deal more than the financial value of the risk avoided. (Otherwise insurance companies would lose money.) In times of flood and sometimes in water supply and irrigation situations, projects may provide safety against unpredictable major losses. As is the case with insurance, this protection provides benefits greater than the money value of the anticipated loss.

The fifth category, not usually considered in a benefit/cost analysis is simply human welfare effects of water resource development. This includes full employment, public health, and education of children, all of which can be affected by projects. The great Fort Peck Dam in Montana is a striking case in point. It was originally started as a work relief project, but there was no way to evaluate this as a tangible benefit. Its primary benefits include hydroelectric power, flood control, and a huge refuge for migratory water fowl, but, initially, during the Great Depression, only the benefits to navigation downstream appeared tangible. The Corps of Engineers had to strain to concoct a favorable benefit/cost ratio, that the districts of Ft. Peck and Omaha were unable to accomplish and was finally done in Washington.

An effect that must be considered, where it exists, is change in the condition of groundwater. Groundwater is sometimes unduly depleted by project withdrawals, or benefited by sustained low flows. Occasionally, the effect upon groundwater is the primary purpose of a project, as indicated in Chapter 4. Effects upon groundwater may be as tangible benefits or environmental intangibles, depending on circumstances.

Sometimes, projects have a significant effect upon water quality. This may be favorable, such as allowing development of a trout fishery just below a dam, or the removal of sedimentation. Variability of flows downstream of power dams may be harmful to fish. If the outlet to a reservoir were designed to release water only from the lower depths of the reservoir, water quality could be adversely affected.

An effect that is often important, especially in the West, is the effect upon the interests, beliefs, and traditions of Native Americans. In building the Columbia River Dam, the tribes originally opposed the program because of the interference with traditional fishing of salmon at the falls, which were to be inundated. There was also the question of native burial grounds that were to be flooded by the reservoirs. These matters required very careful consideration.

Sometimes historical and archeological interests may be involved, or the taking of a church or cemetery.

An objective that should be implicit in the other objectives named is sustainable development. This concept applies to all programs of society, not just to water resources. In general terms, this means essentially taking a long-range view of projects. For example, a flood control reservoir constructed so that it would silt up and be useless after 50 years would violate the principle of sustainable development, even though its B/C ratio for the first 50 years might be favorable.

All of these matters are relevant to the evaluation of a project to the extent that they apply to the particular case, and the concept and design of the project must be adapted to them. Only the first category, the tangible benefits and costs, can be evaluated reliably in dollars and cents, in spite of numerous attempts to quantify environmental values and other intangibles. Other categories can be used in cost comparisons by determining which is the least-cost way of obtaining an important intangible objective. The consideration of all these intangibles is essential to the solution of today's problems. Chapters 7, 10, and 11 indicate some ways to deal with such problems. Although, in principle, intangible benefits are not subject to money evaluation, the Principles and Guidelines (U.S. Water Resources Council, 1983, par. 1.7.2) makes an exception in the case of water-based recreation. It is specifically provided that visitor days of water-based recreation are to be evaluated (in dollars) by the agency.

Government Agencies and Their Functions

Some federal government agencies are very much involved in the problems of the new era, particularly the Environmental Protection Agency (EPA), U.S. Army Corps of Engineers (USACE), the Department of Agriculture (USDA), and the Bureau of Reclamation.

The EPA's first major assignment after passage of the Clean Water Act of 1972 was to control pollution of waterways from outlet works of municipal and major industrial treatment plants. Initially, federal funding was available to defray 75% of the costs of municipal waste treatment plants, and, in general, municipalities were rather leniently treated as compared to industry. It was initially intended that water treatment standards for each area would be decided upon only after a planning process (Section 208), but the process was slow, and soon EPA made it clear that results of such areawide planning would have less force than specific requirements placed directly on the plants. Naturally, after a few years, the 208 process lost its priority and was finally discontinued. Very recently, the EPA has decided that areawide planning would be desirable (See Chapter 8.)

In general, these point-source pollution control programs of the EPA have been successful, and fish are now being restored to some previously contaminated major rivers. More recently the EPA has emphasized a new concept of drainage basin analysis and control. An EPA attempt to prepare a general pollution control program for the Great Lakes made a promising start, but encountered major difficulties (Whipple, 1996, pp. 53-60). The EPA now is initiating a major effort to apply a system of runoff pollution control to industries, municipalities, and selected counties having populations exceeding 100,000 (and not having combined sewer systems). These and other EPA programs relevant to runoff pollution control are covered in Chapter. 8. There is a technical difficulty in coordinating EPA's program with those of other federal agencies, because EPA regions are delineated by state boundaries, rather than river basins.

One of the most powerful (and most controversial) means of preserving the environment is the environmental impact statement (EIS). This is required to be prepared by any agency proposing an action that may have important environmental results (except where a lesser "environmental assessment" is permitted by the regulations). This action is required by the National Environmental Policy Act, under regulations published by the Council on Environmental Quality. The EIS requires an analysis of any adverse environmental effects, alternatives to the proposed action, and any mitigation measures required. The planning effort required is considerable, and any disagreement usually results in litigation (Grigg, 1985, pp. 147-151).

A potentially powerful section of the Clean Water Act is the requirement that the states establish "total maximum daily loads" (TMDLs) of pollutants in rivers, as necessary to obtain desirable levels of water quality. For a long time this provision was generally ignored, because under existing laws, most states do not have sufficient statutory authority to enforce TMDLs on nonpoint-source polluters. However, environmental interests are pushing for action, and within recent years the courts have ruled in Georgia, Idaho, and West Virginia that the TMDLs must be established by the states, or, lacking state action, by the EPA. TMDLs are actually being established in Oregon, although only slowly, with hesitation and difficulties (Anon., April 1997). More recently, by court action, they are now to be required in some parts of California (Anon., June 1997). Obviously, the possibility of TMDLs must be considered in future planning.

In general, the EPA is supported by powerful legislation. Legislative provisions and regulations are often couched in uncompromising terms as absolutes, conveying virtually dictatorial powers to the EPA. Also they are overlapping, so that the same objective may come within several different programs. In practice, the EPA officials usually exercise their great powers with discretion. However, activist environmental interests take advantage of the literal wording of the law and initiate lawsuits that are often successful since judges are bound to enforce the law as it is written. For this reason, the unused provisions of law create uncertainty and make it harder to achieve agreement in solving difficult problems realistically.

The U.S. Army Corps of Engineers, although an integral part of our armed forces, has major responsibilities related to water resources at all times. These functions are exercised by district and division offices, under direction of the Office Chief of Engineers. Each district and division is controlled by a few officers, but the bulk of the staff are civilians. They carry on military construction, but their main mission in peacetime is water resources planning, construction, and management.

At one time, the Corps of Engineers was charged with preparing comprehensive river basin plans throughout the United States (the so-called 308 reports). However, Congress no longer appropriates funds for comprehensive reports by any agency.

Within recent years the Corps has become charged with administration of the wetlands protection statutes, which are now assuming relative importance within the Corps because of the great reduction in construction of new projects.

The Corps administers and operates great numbers of existing reservoir projects—an activity which is normally routine but becomes important at times of large floods or major droughts. The operation of such projects for hydroelectric power may conflict with environmental interests.

Many of the great projects in the West, especially in the Colorado River Basin, were built and are still operated by the U.S. Bureau of Reclamation, mainly for irrigation and hydroelectric power. However, like the Corps of Engineers, the Bureau now seldom gets to build a new major project.

Of other federal agencies, the Department of Agriculture has the closest relationship to the new problems in water resources planning and management, as it controls agricultural operations and the runoff from agricultural land.

The states now have important functions in water development, rivaling the federal departments and central agencies in importance. Other organizations have regional spheres of influence. The TVA is, of course, the prime example. Because of the early success of the TVA, there were various attempts for years to create regional agencies, especially the river basin commissions, that at one time covered many regions of the country. President Eisenhower established a "U.S. Study Commission for Texas," but the idea did not go far. There are still two commissions (for the Delaware and the Susquehanna) in which the United States Secretary of the Interior is a member and full participant in the decisions together with the state governors. There are still a few other commissions in which the states are partners, without a federal member. However, most of the commissions have been abandoned. One of the reasons was the very great measure of control exercised by EPA under the Clean Water Act, which established agency responsibility for functions that otherwise might have been principal concerns of the river basin commissions. However, the commissions, where they still exist, can exercise useful functions of coordination.

The Delaware River Basin Commission is an excellent example. The Delaware River drains parts of New York, New Jersey, Pennsylvania, and Delaware. The commission was created partly to help solve interstate water supply problems, which had resulted in litigation up to the U.S. Supreme Court, and also interstate water pollution problems, which caused great controversy prior to the creation of the EPA. With both the Supreme Court and the EPA having acted, the need is less urgent, but problems remain.

Within recent years the commission has been very active in basin-wide coordination. It established uniform water quality criteria from the head of tide downstream to Delaware Bay (at the request of the three states concerned). It established uniform policies and procedures for waste load allocations and effluent limitations at 83 riverbank wastewater treatment plants, as required to meet the criteria. It created a Toxics Advisory Committee. Monitoring for this toxics investigation revealed, as suspected, that many of the toxics originate in nonpoint sources. The commission developed a new computer model to address water quality issues in the Delaware Estuary (using mainly EPA funds for the purpose) and has chaired a multi-agency CSO task force, the results from which will provide input to the estuary computer model. The commission established the Christina Basin Water Resources Management

Committee to coordinate the policy regarding water quality in this bi-state tributary. Under the commission's drought management plan, a drought warning was issued in 1995 to control releases from storage in the headwaters, but this warning fortunately was ended by heavy rainfall.

This brief summary of only a few principal actions shows that the Delaware River Basin Commission plays a valuable role in coordinating policies between the states concerned and the federal government (Delaware River Basin Commission, 1995). This commission is showing that there are ample opportunities for useful action on a watershed basis. However, in most river basins no such regional agency exists.

One other important federal agency, which no longer exists, is the Water Resources Council. It was created in 1965 to obtain coordination between government agencies, which had been sorely lacking. It was abolished by the Reagan administration (Grigg, 1985). Since that time, comprehensive water resources planning, which the council presumably would have supported, is no longer funded by the Congress. Possible additional uses for a revived Water Resources Council are discussed in Chapter. 11.

References

Anon. "Courts Push States and EPA to Create TMDL Programs," *Environmental Science and Technology*, 31, 4, April 1997.

Anon. "EPA Sets Limits on Non-point Pollution in N. Calif. Waterways," *U.S. Water News*, 14, 6, June 1997.

Delaware River Basin Commission, Annual Report, 1995.

Grigg, Neil. Water Resources Planning, McGraw-Hill, 1985.

U.S. Water Resources Council, "Economic and Environmental Principles and Guidelines," March 10, 1983.

Whipple, W., Jr. *Comprehensive Water Planning and Regulation.* Government Institutes, Inc., Rockville, Md. 1996.

Chapter 2

ENVIRONMENT*

Introduction

A somewhat extreme view of the environmental question (voiced by economists) is the following: "The most serious problem our civilization faces is the ongoing conflict, between economic activity and the biological world upon which all human activity ultimately depends." In attempting to understand this conflict, these economists find that the problem is one of comparing marginal net private benefit in any situation to marginal social costs (including environmental). They discuss the uncertainty in assessing ecological impacts and the fact that the value of biodiversity is unknown. They conclude that a multidisciplinary valuation process is needed to establish goals or targets for environmental protection (Gowdy and O'Hara, 1995).

Although man's attitude toward nature is probably a basic part of our genetic heritage, it must be noted that when men are hard pressed for survival, environmental interests are given little weight. Our remote ancestors—being mostly cold, hungry, and afraid of predators—gave primary importance to food, shelter, and safety from enemies, and, in similar circumstances, so would we. During World War II, the British cut down magnificent beech groves to make tank parks and potato farms. (Some of these groves have now been replanted.) The undeveloped nations cannot establish wild game preserves on a large scale when their populations are short of food. When hard decisions have to be made, many aspects of environmental quality have always given way to necessity. But, in the United States, despite pressing economic and political problems, we still have opportunities to retain many environmental advantages, without costs so great as to impair our ability to support ourselves.

---

* *The environmental matters discussed in this book are those related to water resources. Problems related to energy use and CO_2 emission have worldwide importance, but not for water resource planning.*

Although there is enormous support for preservation of the environment, it is by no means unequivocally clear what the environmental movement means. Some early writers found a religious basis for preserving an undisturbed natural environment. Some later writers, on the contrary, have found in the Judeo-Christian traditions of the Bible a religious sanction for utilizing nature for an expansionist economy (L. Marx, 1966). To most people, a suburban community with trees, shrubbery, gardens, rabbits, the usual variety of little birds, migrating geese and robins, and perhaps a few deer represents a highly desirable environment in which to raise a family, but enthusiasts for the Endangered Species Act consider that some little known plants or animals that have no known connection to humanity are far more important.

Throughout the 20th century, some American leaders have helped preserve the natural beauties of our country, notably through our national, state, and municipal park systems. Later, strong movements appeared to protect all our forests from overexploitation and to protect the soil from erosion and depletion through the Soil Conservation Service. About 1970, there developed a renewed national consensus to protect the environment, especially from pollution. Since that time, the environmental movement has broadened and become more aggressive.

Recently an attempt has been made to unify all the different objectives in the single concept of biodiversity (Takacs, 1996). This attempt brings to light a variety of different ideas. Some people consider the environmental movement to be a natural tendency of mankind resulting from processes of evolution. Some consider that the earth is sacred, and therefore our religion should aim to protect and preserve all living creatures on earth. There is the thought that nature is in an equilibrium, of which all living creatures are a part, and that to disturb this equilibrium by loss of any species would lead to the risk of destruction of all earth's creatures, including mankind. Others consider that living creatures, especially plants, through their undiscovered qualities and genetic characteristics, have huge potential for economic and human health benefits, which would be jeopardized if all species were not preserved. Still others consider that the artistic beauty of undisturbed nature, as revealed particularly through discovery of rare species, is a major value of biodiversity.

It is astonishing that such diverse ideas as the above could be advanced by intelligent people who supposedly have a common objective in mind.

Some of these ideas are obviously wrong. For example, the idea of an equilibrium of species in nature is completely negated by the history of evolution on this earth, where plants and animals have competed for survival, and great numbers of species have appeared and disappeared without any generally catastrophic consequences. [About 230 million years ago, approximately 90% of marine species disappeared (Gowdy and O'Hara, 1995, p. 166)].

The artistic beauty of undisturbed nature and the alleged genetic love of humans for biodiversity are both far from universal. Some natural species are definitely ugly, while others, like snakes and many insects, are naturally repulsive to humans. Some, like the zebra mussel, are harmful to other species considered valuable to man. To a biologist, the discovery of a new species of plant or insect is understandably an intellectual thrill, but such feelings cannot have been inherited in the genetic disposition of mankind in, general. Our primitive ancestors were simply incapable of understanding such fine distinctions. What has been inherited in most persons is a love of (some) animals, particularly young animals, and enjoyment of open space, fresh air, and such natural beauties as the sunset. A more recent analysis of biodiversity and ecological integrity explains these concepts further, but it also stresses the manifold difficulties in applying them, and the many errors in the attempts to do so (Frissell and Bayles, 1996).

There is no doubt that there are close relationships between species and that in some way a composite of species constitutes value in the environment. Judging the Columbia River entirely by salmon, the Rapid City streams entirely by trout, and all streams by endangered species is only the crudest of approximations, and ultimately we should come to something better. However, in the meantime, the number of species lost certainly gives ground for thought. It has been estimated that the loss in biodiversity due to human activity since the Industrial Revolution is 10-20% (Wilson, 1992). Whatever credence we give to these numbers, there is no doubt that the trend is undesirable.

Another new concept in the understanding of the environment is dynamic equilibrium. Although nature has often been portrayed as being in stable harmony prior to the advent of man, this is not really the case. It has long been realized, in the field of biology, that predator-prey relationships often result in periodic fluctuations, whereby, over a period of years, the predator flourishes and multiplies for several years until the prey become scarce, after which there is a decline in the number of predators, and the cycle is repeated. It is now pointed out that other types of dynamic equilibrium exist in nature, particularly in the growth of forests. Fires periodically destroy large parts of a forest, which then gradually grow back again. These changes have corresponding effects upon the growth of vegetation, wildlife, and even salmon (See Chapter 6). Although the various species have some natural resilience to cope with such changes, it is important that man-made changes, such as the harvesting of lumber in forests, be planned so that this resilience, over a period of time, can restore the damage done (Huntsinger, 1997).

Besides the instinctive reactions of mankind to nature, most thoughtful people support the basic idea of conserving those attributes of nature that benefit mankind, either now or in the future. This thought does not require religious sanction. What it does require is a scientific basis. It requires knowledge of broad worldwide problems such as those linking the destruction of tropical forests and the spreading of harmful chemicals from industry to world climate and weather. It requires knowledge of which among the many species of plants may be actually useful and which really harmful. For example, the spreading of vegetation along streams is generally beneficial to wildlife, but phreatophytes may utilize water to no purpose in arid regions, and inedible sagebrush may crowd out more useful vegetation in cattle country.

Sometimes the environmental war-cry is raised to protect quite different interests. Some years ago, long-range needs of New Jersey and Pennsylvania for flood control and water supply were proposed to be met by construction of the Tocks Island Dam on the Delaware River. Proponents noted that it would also provide a fine recreation spot for urban inhabitants of New York City, Newark, and Jersey City. A wave of environmental objections soon arose, the governor of New Jersey withdrew his support, and the project had to be abandoned. However, a considerable part of the objections stemmed from northwest New Jersey, which is a lovely countryside of quiet homes and villages. Apparently, the inhabitants were apprehensive of hordes of undesirable visitors from the central cities. Such views are "politically incorrect" and a long way from the high ideals that should distinguish environmentalism.

Wetlands Control and Stream Restoration

Through their intermittent flooding and great extent, wetlands consti-
tute a major aspect of the hydrologic cycle of rivers and estuaries. From
an environmental viewpoint, they are of prime importance as wildlife
habitat. Originally, "swamps," as wetlands were then called, were viewed
mainly as an obstruction to agriculture and other development. Consequently, by the mid-1980s, the nation had lost over half of its original
wetlands (Teels, 1994). This was accompanied by corresponding reductions in birds, fish, and wildlife. It gradually came to be realized that
wetlands were very important as ecosystem habitat, and federal wetlands protection began with Section 404 of the Clean Water Act, which
requires permits for various actions including draining wetlands.

Wetlands may be harmed by the programs of the construction agencies
mainly in three ways. (1) Wetlands frequently are used to construct reservoirs of any size. (2) Flow regimens modified to provide maximum
advantage for various purposes, including flood control, navigation or
power production, may adversely affect wetlands, or, less frequently,
may, expand them. (3) Channel improvements and levees built in the
interest of flood control may have the result of draining wetlands. However, wetlands may also be created or restored by construction, as will
be examined later.

The national program of wetlands preservation is administered by the
Corps of Engineers, the Department of Agriculture, and the EPA
(Whipple, *et al.*, 1994). Under a 1994 memorandum of agreement between the Army, the Environmental Protection Agency, the departments
of Agriculture, and the Interior (*Federal Register*, Jan. 19, 1994), the Soil
Conservation Service is the lead agency for wetlands delineation on all
agricultural lands. The Corps of Engineers implements wetlands control on other lands but is subject to a measure of control by the EPA and
the Fish and Wildlife Service. Losses of wetlands of less than 1/3 of an
acre do not require permits, and those of three acres or less are
approvable by an expedited procedure. However, at best, the approval
process for permits is so cumbersome that many developers or municipalities, when informed that a wetlands permit will be required, simply
look for another site.

Wetlands permit requirements create major obstacles when construction of regional stormwater detention basins are proposed. Systematic

planning of stormwater detention on a watershed basis (Chapter 9) usually requires regional control impoundments (rather than only those at
site). Such structures almost always include wetlands in the area to be
flooded. In fact, opportunities may exist to utilize an existing wetlands
as a detention basin, to the advantage of both purposes. Institutional
problems make such arrangements difficult. However, unless the existing wetlands has some particular advantage, such as benefit to an endangered species, the combination of wetlands and stormwater management should be favored.

In many cases, especially where important highways are involved, adverse wetlands impacts can be offset by creating or improving wetlands
elsewhere. Construction of new wetlands requires careful study of the
existing hydrology and the possibility of using the wetlands to help control stormwater flows and to improve the quality of the runoff, as well as
to provide the ecological advantages that are usually the basis for the
project.

The President's announced "interim" policy is that no additional wetlands should be lost. In view of the necessary use of some wetlands for
indispensable purposes such as major highways, and the annual loss of
some wetlands through sedimentation in stream valleys and deltas, this
policy of no net loss requires the restoration of many lost wetlands.
Moreover, prominent environmental interests urge a major restoration
of lost wetlands.

The plans for restoring lost and damaged wetlands in southern Florida
are large and complex, but have a good chance of being effective
(Appelbaum, 1998). The Central and Southern Florida project of the
Corps of Engineers was designed to provide flood control, water supply
for many purposes (including agriculture), prevention of saltwater intrusion, and protection of fish and wildlife. It has fulfill,ed its objectives,
but has had unintended adverse consequences on the Everglades and
Florida Bay ecosystems. In 1992, Congress authorized the Corps of
Engineers to undertake a comprehensive review to determine the feasibility of restoring the South Florida ecosystem while meeting other
needs. The Corps undertook this study in partnership with the South
Florida Water Management District.

In 1994 the governor established the Governor's Commission for a Sustainable South Florida to recommend strategies ensuring the long-term
compatibility of a strong South Florida economy and a healthy South

Florida ecosystem. This commission, 47 members representing various interests and groupings of society, has prepared a conceptual plan of the restudy to be used as a starting point by the restudy team. The Corps and the Commission appear so far to be working in harmony, and, since only one state is involved, and protecting the Everglades is a clear objective, there seem to be good prospects of a successful outcome.

Restoration of lost wetlands is usually undertaken to compensate for those lost through a particular project. It can pose a major obstacle in some cases. For example, the state of New Jersey was about to build a much-needed major water supply project, the Manasquan Dam, when, suddenly, after years of prior planning it was brought out that some of the land inundated by the reservoir would be wetlands. Nothing could be done until the state agency, by a stroke of luck, was able to obtain from another agency the transfer of lands that could be used to create additional wetlands, thus compensating for the loss without adding to the cost of the project.

The Corps of Engineers has initiated a substantial program of creating wetlands using dredged material (Landin and Patin, 1993). The material is available in quantity, mostly from routine maintenance of navigation channels, and it must be disposed of somehow. The need for such arrangements is greater near urban areas, where disposition would otherwise be a problem. This approach has been used in a number of cases to create wetlands, some of large size. These wetlands have value as habitat, but they also may provide for other planned uses, such as park facilities, visitor centers, nature trails, biking, hiking, bird watching, and boating.

An entirely different type of re-creation of wetlands was involved in the "Brandon Town Center" in Florida (Miller and Feher, 1993). In this case, an area mostly in the floodplain of a creek was developed partly for flood control purposes and to provide for typical urban development, but also to provide a desirable wetlands environment especially adapted to wading birds. Prior to the project, the area had been taken by vegetable and sod farming and a cattle ranch. Since project construction, 39% of the area has been dedicated to surface water management and wetlands preservation.

In view of the importance of conservation and restoration of wetlands, attention is being given to design and planning (Hayes, *et al.*, 1993).

Functions performed by a wetlands may include groundwater recharge, water quality improvement, flood control, and recreation, as well as fish, wildlife, and vegetative habitat. A number of design requirements, and construction techniques have been developed, but the various functions of a wetlands are not separable but closely interrelated, and vary widely from site to site. The usefulness of wetlands depends upon nutrient availability, soil conditions, water depths, water chemistry, flow conditions, and types of vegetation. Useful wetlands include such diverse types as southern swamps, prairie potholes, coastal marshes, and mountain pools.

Wetlands in the form of natural marshes can be utilized for stormwater management with some success. A substantial reduction in pollutants has been achieved in experimental projects (Rushton and Carr, 1993). Much more frequent is the creation of new wet ponds and wetlands as part of detention basins built for stormwater management. This is standard technique in the modern stormwater management systems designed to improve water quality. The detailed techniques of using stormwater impoundments to provide the advantages of wetlands are specified in an excellent practice manual (Schueler, 1987). Design requirements are given under the heading "Shallow marsh creation." The main point is that planting aquatic vegetation around the perimeter of a wet pond protects the bank from erosion as well as encouraging fish and wildlife.

In northwest Georgia, construction of a water supply reservoir was made contingent on the replacement of 29 acres of wetlands filled or flooded by the reservoir (Reeves, 1996). It was originally suggested that replacement be on a two-for-one basis. A much more complex plan was finally developed to preserve considerable areas of upland forest and bottomland hardwood swamp, to enhance 30 acres of bottomland swamp and wetland, and to provide wetland hydrology to 13 acres of cleared floodplain. Careful planning was essential.

As an exceptional situation, in Yolo County, California, citizens obtained over $15 million to create the Yolo Rypass State Wildlife Area, a wetlands area located on the Pacific Flyway bird migration route (*U.S. Water News*, 1996).

Artificial wetlands may sometimes be used as one aspect of wastewater treatment. However, this technique is not well developed, and should be considered with caution.

The situation concerning the control of wetlands is not satisfactory to environmentalists, who would like not only to stop any new losses of wetlands but also to restore (much of) the wetlands already lost. To developers and growing municipalities, wetlands control is an obstruction to necessary improvements. From a scientific viewpoint there continues to be great uncertainty. The ecological value of various kinds of wetlands obviously varies greatly. Neither the Corps of Engineers nor the Department of Agriculture is an environmental organization fully staffed to answer such questions. Despite the scientific uncertainties, the program is a necessary one. However, it appears capable of improvement.

In many areas, channel modification, urbanization, or unusually severe floods and erosion have destroyed much of a stream's habitat value. Recently, increased attention has been given to restoration of modified or otherwise degraded streams (Rinaldi and Johnson, 1997). Techniques of restoration generally include (a) construction of asymmetrical cross-sections, (b) inducing the stream to develop point bars, (c) two-stage channel designs, (d) floodplain approaches, (e) pools and riffles, (f) re-creation, and (g) sinuosity and meander restoration. If meander restoration is to be attempted, it is important that careful analysis be undertaken in order to determine the appropriate geometry and meander size.

Limitation of bank erosion has usually been done by stone riprap, but stabilization by live willow posts may be effective, while also providing environmental advantages (Watson, et al., 1997).

Rapid City, South Dakota, suffered a disastrous flood in 1972. After the flood, the channel lacked the pools, riffles, and overhanging banks that normally provided fish habitat. A multipurpose plan was needed (Peters, *et al.*, 1995). The Corps of Engineers formed a partnership with the city and the state's Department of Game, Fish, and Parks to restore more than nine miles of the stream, through development of a low-flow channel as well as a flood-capacity channel. The low-flow channel incorporated flow deflectors, pools and riffles, overhanging vegetation and ledges, boulders, snags, simulated drops, and other habitat features. The project was judged successful because it restored the trout population, and reduced the white fish population. (It may be noted that this favorable judgment does not attempt to consider advantages of aquatic biodiversity.) With this sort of cooperative approach the best plan for the community can be achieved.

The disastrous Missouri River flood of 1993 indicated that woody corridors in the floodplain behind levees have a significant effect in protecting the levees from breaching by the floodwaters (Dwyer, *et al.*, 1997). Although these wooded corridors in an otherwise developed floodplain were recommended from a flood control viewpoint, the presence of these extensive woodlands rather than agriculture have the favorable environmental effects of increasing wildlife habitat and re-establishing natural riparian ecosystems.

Parks and Constructed Environments

Parks and constructed environments are designed to preserve attractive portions of our natural environment, which they generally do successfully. There are of course limitations imposed by the numbers of visitors. It is not necessary for environment to be natural in order for it

Figure 2-1. Wildfowl on Potholes Reservoir. Courtesy of U.S. Department of the Interior, Bureau of Reclamation.

to be valuable and useful. Figure 2-1 shows a beautiful view of waterfowl in a western reservoir built in arid country near national flyways. It is the Potholes Reservoir, a 25,000-acre body of water, and it has been estimated that up to 125,000 ducks have occupied the reservoir.

New York City's Central Park illustrates the very great public use of open space, bringing the pleasure of outdoors to millions of crowded inhabitants of that great city. Obviously, however far it is from its natural condition, that park has environmental value to a great many people. In another extreme example, the "petting" zoo brings a touch of nature to urban children.

The national policy on wetlands in substance illustrates the value of constructed environments because it allows wetlands to be taken for development, with the agreement to build new wetlands in exchange.

Within recent years, besides the parks and forests, a number of ecosystem management areas have been created (Yaffee, *et al.*, 1996). A total of 105 are featured in this reference, in all parts of the country. It is interesting to note that, of the 14 principal ecosystem stresses to which these areas are subject, only one, "hydrologic alteration," seems likely to result from water resource development. No one should get the idea that water resource development and ecosystems are mutually opposing. In fact, water resource development favors ecological values as often as it harms them.

Water Quality

The great national movement that culminated in the Clean Water Act of 1972 was aimed primarily at improving water quality, initially at point sources of pollution. The relevant EPA programs are covered in Chapter 8.

Environmental Effects Downstream of Dams

Although the environmental effect above dams in reservoirs is predominantly favorable, downstream changes may be unfavorable. A recent survey summarizes: "Several issues take the forefront in consideration of adverse effects of dam operations. Native fish, protected under the

Endangered Species Act, are thought to be threatened by the clear, and usually cold, releases from dams. Riparian vegetation can either be enhanced or degraded by dam operations. Streamside and channel sedimentary deposits are critical; too much sediment can aggrade channels and cause flooding problems, whereas erosion of sediment can degrade habitat and decrease the potential for recreational use" (U.S. Geological Survey, 1996).

For salmonids, there are other adverse environmental conditions associated with dams, as explained in Chapter 6.

Endangered Species

One of the most powerful environmental principles in the United States is the protection of endangered species of animals and plants and the preservation of their habitats. The basic principle is that all plants and animals are important to mankind, even though the relationship may not be known, and that it is harmful to us to allow any species to disappear completely. This view is reinforced by specific instances in which it has been found that previously inconspicuous plants or animals provide life-preserving substances or improvements to genetics of commercially valuable species. However, this argument is weak because of the disappearance in the past of vast numbers of species (with no known dramatic consequences, as indicated earlier in this chapter) and because of the obviously adverse effects of some species, such as the anopheles mosquito, and the AIDS virus. The biological state of the world is not fixed but variable, with constant competition between species and new species of both plants and animals evolving to fill any gap

In the United States, the preservation of endangered or threatened species has been mandated by the Endangered Species Act. (It is interesting to recall that in Great Britain somewhat similar protection was provided for some species in the 19th century. Much earlier, in England, the swans on the Thames River were considered property of the King, and therefore were protected, under penalty of death.) Under the Endangered Species Act, control is based upon federal lists of endangered or threatened plants and animals. Enforcement is implemented through various federal programs such as permits to encroach upon wetlands, and requirements for environmental impact statements.

Although the protection of endangered species is mandated as being beneficial to mankind, there is no provision in the law for counterbalancing the adverse impacts upon endangered species with other losses to mankind that may result from the preservation of the endangered species. Protection of endangered and threatened species is the law. That this is a desirable national objective there is no doubt. However, the extent to which this principle should be maintained as an absolute is considered in the following section.

Relative and Absolute Values

Many attempts have been made to evaluate environmental values on a money basis (Woodward and Huppert, 1993). Such evaluation would be very convenient for computer analysis of systems. Sometimes approximations are made on the basis of expenditures of individuals while hunting or fishing, or of demand value for certain privileged recreational opportunities. Wetlands may be evaluated by the value of flood losses prevented. The most ambitious attempt to establish such a system of values concluded with a rough estimate that the total worth of environmental aspects of the biosphere was 33 *trillion* dollars (*Science*, 1997). It is notable that in this estimate wetlands and coastal areas provided just over half of the entire global value. However, such values have not been established with sufficient plausibility to be applicable for water resource planning purposes.

We should acknowledge the idea that the various measures for protection of the environment cannot be absolutes. With the continuing increase of population in this country, space must be found for more people every year. Although, within the foreseeable future, there will always be open space, parks, forests, and mountains in the United States, many of our urban populations live in overcrowded, unsanitary conditions and need more space to raise families. There are good reasons to protect the environment, but this objective must be reconciled with the need to provide a healthful life for human beings. We must consider the other basic public objectives summarized in Chapter 1. Worthwhile public objectives that may be affected by productive water resources development include low-cost housing, public health and welfare, full employment for all, good education for all children, and care of the aged. Although our Constitution assures "life, liberty and the pursuit of happiness," even these are not absolutes. Compromises are made between various public objectives. For example, convicted criminals have a much-restricted basis for enjoying liberty. If protection of some "endangered" species causes thousands of people to lose their jobs, this may be in accordance with law, but still may not be automatically the best course of action. There should be an opportunity to determine whether the environmental loss due to destruction of the species' habitat and gene pool really warrants the adverse economic consequences if the habitat is to be preserved intact. In many cases, a compromise solution may be developed that would preserve or replace sufficient essential habitats, while allowing economic activity to continue. Or it may be that some endangered species should be allowed to disappear, like the dinosaur, the mammoth, and the virus that caused the worldwide influenza epidemic of 1918. Environmental objectives are important and should be protected by law, but the laws and institutions should require that ultimate decisions be made with consideration of all national objectives and not only one.

In the meantime, planners must face the uncertainties and frustrations of the Endangered Species Act and its effects, through wetlands protection programs and the environmental impact statements. Some of the problems of doing this in specific cases are outlined in Chapterter 10.

References

Appelbaum, S.J. "Consensus Based Planning: Developing a Conceptual Plan for Everglades Restoration." Session on the Wetlands, Conference Proceedings: Coordination between Water Resources Planning and Environmental Regulation, ASCE, 1998.

Dwyer, J.P., Wallace, D., and Larsen, D.R. "Value of Woody River Corridors in Levee Protection Along the Missouri River in 1993." *Journal American Water Resources Association*, 33, 2, 481, April 1997.

Frissell, C.A., and Bayles, D. "Ecosystem Management and Conservation of Aquatic Biodiversity and Ecological Integrity." *Water Resources Bulletin* 32, 2, 229, April 1996.

Gowdy, T. and O'Hara. S. Economic Theory and Environmentalists. St. Lucie Press, 1995.

Hayes, D.F., Crockett, T.A., and Arends, M.T. "Wetland Engineering, Design, and Construction." In *Water Management in the '90's*, ASCE, 1993, p. 88.

Heimlich, R.E. "Wetland Policies in the Clean Water Act," in "Water Resources Update," Universities Council on Water Resources, No. 94, Winter 1994, p. 52.

Huntsinger, L. "Managing Nature: Stories of Dynamic Equilibrium," in "What Is Watershed Stability," Water Resources Center Report No. 92, University of California, April 1997.

Landin, M.C. and Patin, T.R. "Innovative Wetlands in Urban Settings, Using Dredged Materials." In *Water Management in the '90's*, ASCE, 1993, p. 109.

Marx, Leo. "American Institutions and Ecological Ideals," *Science*, Nov. 27, 1966, p. 945, vol. 171.

Miller, C.L. and Feher, G.G. "Innovative Wetlands Re-creation in Urban Settings." In *Water Management in the '90's*, ASCE, 1993, p. 113.

Peters, M.R., et al. "Assessment of Restored Riverine Habitat Using RCH HARC." *Water Resources Bulletin*, 31, 4, 745, Aug. 1995.

Reeves, Jim. "Anatomy of a Wetland," *Civil Engineering*, Jan. 1996, p. 58.

Ruston, B. and Carr, D. "An Isolated Wetland Used for Storm Water Treatment." In *Water Management in the '90's*, ASCE, 1993.

Schueler, T.R. "Controlling Urban Runoff." Washington Metropolitan Water Resources Planning Board, 1987, p. 9.16.

Takacs, D. *The Idea of Biodiversity*, Johns Hopkins University Press, 1996.

Teels, B.M. "Agriculture and Soil Conservation" in "National Water Resources Regulation: Where is the Pendulum Now?" ASCE, 1994.

U.S. Geological Survey. "Dams and Rivers: Primer on the Downstream Effects of Dams." Collier, M., Webb, R.H., and Schmidt, J.C. Circular 1126, 1996.

U.S. Water News, "California Wetlands Project Thrives Through Public/Private Partnership," Dec. 1996.

Watson, C.C., Abt., S.R., and Derrick, D. "Willow Posts Bank Stabilization." *Journal American Water Resources Association*, 33, 2, 293, April, 1997.

Whipple, W., Jr., et al. In: "National Water Resources Regulation: Where is the Pendulum Now?" American Society of Civil Engineers, 1994, p. 258.

Wilson, E.O. *The Diversity of Life*. Harvard University Press, 1992.

Woodward, E.J. and Huppert, D.D. "Role of Economics in Endangered Species Act Activities Related to Snake River Salmon," in "Water

Management in the '90s," ASCE, 1993.

Yaffee, S.L., *et al. Ecosystem Management in the United States.* Island
Press, Washington, D.C., 1996.

Chapter 3

POPULATION AND DEVELOPMENT

Introduction

In order to understand the nature of the economic aspects influencing water resources development in the United States, it is necessary to consider population growth not only in the United States, but worldwide.

Birth Rate, Immigration, and the World Picture

According to the *World Almanac, 1997*, world population increased from 1980 to 1996 and is estimated to increase in the near future in accordance with the following table.

Year	Population (in Millions)	Increase (in Millions)
1980	4,458.0	
		>1,313.9
1996	5,771.9	
		>1,090.8
2010	6,862.7	

Accordingly, world population has recently increased by about 81 million annually. In the period 1990-1995 the United States population increased from 248.7 million to 265.6 million, a little over 1% annually. This increase in the total U.S. population is not only from births within the country, as there is a considerable amount of both legal and illegal immigration across our borders. The latest projections from the United Nations show a marked reduction in the rate of increase but confirm that world population remains a serious problem (United Nations, 1997).

From the viewpoint of water resources, this situation means that planning for the future in the United States must envisage a continued increase in population growth, much of it from abroad. (The world population situation is also of very great significance from some other viewpoints, but they are outside the scope of this book.)

Urbanization, Industry, and Deterioration of Large Cities

As our population increases, it is becoming more urbanized. The center cities have a large resident population, but each is surrounded by a periphery of smaller towns and suburbia, where the more prosperous members of the population mostly live, especially those with children. Despite higher rents per acre, the central cities provide homes for most of the poor. Living conditions are crowded and unsanitary compared to conditions prevalent in suburbia. Large cities are featured by financial, legal, cultural, and administrative activities, but the bulk of the physical manufacturing takes place elsewhere, particularly in areas where wage rates are lower. Despite centralized sewer systems, cities originate a lot of water pollution, from street runoff, roofs, filling stations, and illicit or accidental discharges. Unlike the surrounding suburbs, cities lack the space to provide effective remedial means. (See Chapter 9.)

Encroachment upon Open Space and Agriculture

With the increase in population and the corresponding demand for housing (in other than urban environments), there is a continuous taking of open space and agricultural land for housing. Of course, there is some construction of individual houses, but most of the taking is in the form of developments. There is still some addition of land for industrial purposes, frequently with acreage for prestige purposes, or reserved for future expansion. A considerable amount of both residential and industrial development occurs in floodplains.

Nonpoint-Source Pollution

In the United States, the growth of physical manufacturing is probably less than the growth of population because articles sold are becoming more sophisticated. The larger manufacturing establishments have long had their liquid waste outputs subject to control under the Clean Water Act. However, pollution from their runoff is still largely uncontrolled. Runoff from agriculture is largely controlled for some purposes by the Soil Conservation Service. The usual control of erosion largely reduces sedimentation, but the increasing use of fertilizers and pesticides is adding to pollution. The more flagrant cases of large-scale mining pollution have been largely controlled, but many small-scale operations are still polluting the waters. Above all, the continued increase in homes and private automobiles, resulting from the increased population, provides steady increases in pollution from driveways, garbage, home gardens, lawn fertilizers, and miscellaneous spills. Although the owners are not aware of it, even a well-kept suburban home with a good sewerage system adds to runoff pollution.

Conclusion

With continued population growth, and with institutions and laws as they now exist, the United States must face a general increase in water supply demand. There will also be increased flood control needs and increased pollution from nonpoint sources. There will be increasing pressure on the environment. Such needs are to be expected and must be provided for. (In many foreign countries, population is growing more rapidly, and accompanying problems, especially supply of water and pollution, will be more acute than in the United States.)

References

United Nations. "World Population Prospects, 1996 Revision." 1997.
World Almanac, 1997.

Chapter 4

WATER SUPPLY

Introduction

Of all of the uses made of water, water supply is the most ancient, as water was used for family purposes even in prehistoric times. However, the federal government's interest in water supply is fairly recent. Although water for irrigation in the West became important fairly early, the incorporation of water supply as a purpose of federal multiple-purpose projects was not generally authorized until the Flood Control Act of 1936. Earlier in our history, the provision of water, other than for irrigation projects, was considered to be the business of individuals, businesses, and municipalities, under supervision by the states rather than the federal government. From society's point of view, the supply of drinking water is absolutely indispensable, more so than any other use of water. However, within the category of water supply, there is much low-priority usage, as will be discussed later in this chapter.

In most river basins, the responsibility for planning water supply is not clear-cut. In some cases, federal reservoirs may be the key to water supply availability, and there are always other federal interests and responsibilities such as maintaining water quality and fish life, preserving wetlands, and navigation. On the other hand, most states assume at least some responsibility for water supply planning, and many municipalities and privately owned utilities have major water supply operations. Moreover, in time of drought, state/local demand management programs are a necessity. Although a federal project may be built with water supply as a major purpose, non-federal interests often pay the costs. Therefore, it is usually not clear just who is responsible for overall water supply planning.

A considerable proportion of the nation's water supply comes from groundwater. Traditionally, water from wells was considered pure, and it has been realized only within recent years that much well water is contaminated with volatile organics and other pollutants. Concentrations of these substances, imperceptible to the eye and tasteless, may be carcinogenic or otherwise harmful. Moreover, in addition to the pollutants currently found in well water, there are numberless places throughout the country where groundwater has been contaminated, but the slowly moving plume of contamination has not yet reached the wells. In other cases, such as the famous Ogallala Aquifer, the mass of water underground is being steadily depleted, with no recharge in sight.

The Safe Drinking Water Act is increasingly raising problems for water systems in the form of many new substances that must be tested for, and criteria that preclude the use of some water supplies or require new forms of treatment. (See Chapter 8.)

Some of the new priorities, especially requirements for low flows for preservation of fish life and for other environmental purposes, run counter to long-established systems of water rights in the Western states.

Value of Benefits

In federal water planning, projects are generally evaluated by a benefit/cost ratio. However, recognizing that the continuity of water supply is essential to the survival of a community, an important qualification is made for water supply. The federal conceptual basis for evaluating water for municipal and industrial purposes is the willingness to pay. Where price reflects marginal cost, price may be used. If such measures are not available, the resource cost of the most likely alternative may be used (U.S. Water Resources Council, 1983). This is not strictly correct in principle as obviously there is some limit to the feasibility of supplying water in remote locations. However, this discrepancy is not a real obstacle. The benefit/cost ratio is seldom the determining factor for the feasibility of proposed new water supply projects.

Rivers, Lakes, and Reservoirs

It is generally accepted that natural rivers, streams, and lakes are available for water supply, but this assumption is subject to many qualifications. In the West, of course, water may be taken only when allocated by water rights. In granting water rights, preference is usually given to private and municipal uses, but severe problems may arise when large new developments are contemplated in an agricultural community where crops are irrigated with limited supplies of water. In the East, under original common law, lands abutting the river (riparian lands) could use water within reason, but many states now allocate water by permit, so that competition arises between different uses of water, similar in principle to competition in the water-rights states. In each case, the preferred solution is usually to increase the reliability of the available supply, either by structural means such as reservoirs or wells, or by moving water from other more distant sources. The backup approach to assure meeting the highest priority needs is usually water conservation. All of these alternatives are discussed later in this chapter.

The use of water from natural waterways is subject to various restrictions. The most important difficulties arise because of water quality. Natural water is almost always required to be treated prior to use, with the extent and kind of treatment varying widely. The main programs of water pollution control that are required for broad environmental purposes are designed as to reduce pollution from municipal waste treatment plants and major industries (Chapter 8). These programs are generally successful, but they have hardly made a start in reducing nonpoint sources and other dispersed sources of pollution.

The estuaries of large rivers are tidal, and saltwater flows in and out with the tides, especially during drought. This not only makes larger portions of the river unusable for drinking water during such times, but also may contaminate groundwater nearby. The supply of fresh water on coastal islands is a problem, especially if they become heavily developed.

Rivers are of course subject to floods. (See Chapter 5.) Water supply intakes must be built to be functional at high water, excluding floating debris. At such times, the water treatment plants must be prepared to handle large loads of silt, oils, salts, minerals, and other contaminants flushed from the land, and vegetable detritus. Also, being close to the river, the water treatment plants should be protected from floods. Water intakes must also be designed to withstand freezing and floating ice.

Groundwater and Its Conjunctive Use

Groundwater is very important in the United States. In 1985 it was estimated that 34% of water withdrawals from any source (excluding hydroelectric power) were from groundwater, that 51% of our population was served by groundwater, and that 40% of all public water supply came from groundwater (Solley, 1988). The combining of water supply from surface sources with that from wells (a process known as conjuctive use) has considerable advantages. The capacity of a groundwater supply is limited not only by the size of the wells and pumps, but also by the water situation underground. Usually, the underground aquifer from which water is drawn has limited recharge capacity, but it may be pumped at a higher rate for some weeks or months at a time. This may provide water necessary to meet peaks in demand, such as water for irrigation, or to accommodate temporary reductions in surface water capacity caused by such events as large spills of polluted material, equipment outages, or floods. Conjunctive use provides for a more flexible operation, instead of requiring storage or connection between adjacent water supply systems.

Conjunctive use must be distinguished from simple overuse or "mining" water from aquifers in excess of their natural recharge. An abuse of this sort is that of the Ogallala Aquifer in western Oklahoma and north Texas, which has been mined for many years. This is leading inevitably to a bad situation, as there is no other source of water nearby. Overpumping of the aquifer in Albuquerque is not only exhausting the available water supply but, by settling of the soil, even damaging structures (*New York Times*, 1997). In a much smaller, earlier case in California, two entrepreneurs enriched themselves by growing irrigated long-staple cotton using illegal Mexican labor, while the groundwater level for their wells sank 800 feet below the ground, greatly to the detriment of other users.

Deep wells usually provide relatively pure water, which is easy to treat. Shallow wells are much more subject to contamination. Industrial contamination is very important. Some agricultural activities, such as feed lots, are exceptionally likely to pollute the groundwater, which then flows into an aquifer and may be drawn into a well. A very common source of home well pollution is the septic tank. The author recalls a young relative occupying a rented house with a well and a septic tank. She complained that her well water foamed a day or so after she had washed her clothes! In one notorious case in New Jersey, several hundred homes were built on small lots, each with its own shallow well and septic tank. Inevitably, each well was soon contaminated. After the resulting public uproar, action was taken by the state to prevent such occurrences. However, it must be remembered that substitution of conventional sewerage systems for septic tanks, while avoiding pollution, results in depletion of groundwater.

The pollution of groundwater is not due only to installations such as septic tanks and cattle feed lots. Nitrates in fertilizer, used on farms and also on home lawns and gardens, can seep into groundwater, which can be very harmful to pregnant women. It can also produce the "blue baby syndrome." Also, nutrients cause excessive algal growth in lakes and estuaries. Deposits of hydrocarbons from automobile traffic enter run-off and cause pollution, especially after chlorination. Pesticides are harmful in many cases. Along the Eastern seaboard, groundwater may be invaded by infiltrating sea water where excess withdrawals from an aquifer lowers pressure. In some areas, such as southern New Jersey, there are extensive areas where sea water penetrated thousands or even millions of years ago. This is known as connate saltwater. Groundwater may also be polluted by outflows from polluted rivers and streams or saline estuaries, in cases where the groundwater has been depleted to an unusually low level.

As urbanization continues in areas with permeable soils, waste disposal and storage of chemicals are apt to pollute the groundwater over a period of time. One technique of minimizing such pollution is to establish well-head protection areas around each well. In principle, pollution-producing activities should be excluded from such areas. These areas should not be circular but should instead allow for the direction of groundwater flow.

Treatment, Including Ozonation

Until recently, treatment of potable water meant only chlorinating, clarifying, and filtering the raw water. These steps are still necessary and filtration is essential. The EPA recently sued New York City for failing to provide filtration for water from reservoirs in the Croton Watershed (Revkin, 1997). Chlorination kills bacteria but reacts with certain waters to produce harmful by-products, notably trihalomethane (THM). To cope with such situations, ozone can replace chlorine as the primary treatment agent, but a small quantity of chlorine is added after ozonation to maintain water safety and to provide a convenient means of measuring the degree of safety as it passes through the distribution system. Ozonation is costly, but it provides a means of meeting standards of the Safe Drinking Water Act, with waters that would otherwise be unusable.

Ozonation is used in the Tri-County Water Supply Project of the New Jersey-American Water Company. The raw water used is from the mainstem Delaware River, just above Camden and Philadelphia, and the quality of the water is such that it would be extremely difficult to meet water quality standards with conventional technology. The plant's capacity is 30 million gallons per day (mgd) and is expandable to 60 mgd. The raw water pumping station is designed to bring the water from the river bank to a more suitable location for the treatment plant.

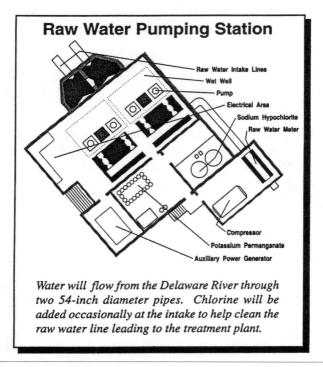

Raw Water Pumping Station

Raw Water Intake Lines
Wet Well
Pump
Electrical Area
Sodium Hypochlorite
Raw Water Meter
Compressor
Potassium Permanganate
Auxiliary Power Generator

Water will flow from the Delaware River through two 54-inch diameter pipes. Chlorine will be added occasionally at the intake to help clean the raw water line leading to the treatment plant.

Figure 4-1. Raw Water Pumping Station. Source: New Jersey American Water Co.

The ozonation system will include liquid oxygen storage, ozone generation, contact chambers, and off-gas treatment. Successive stages are as
follows:

- Intakes

- Raw-water pump station

- Raw-water reservoir

- Ozone contactors

- Superpulsators (for rapid clarification and settlement)

- Filters

- High-service pump station

- Addition of low concentration of chlorine

- Distribution system

The plant uses ozone generated from liquid oxygen, in combination with
superpulsators for high-rate clarification. Deep-bed, micromedia carbon
filters are used, as they are biologically active and consume the man-
made chemicals present in the water. Advanced automation and instru-
mentation are used so that the entire plant can be operated with a staff
of four. (See Fig. 4-2.)

Reprocessing Wastewater

Sometimes reclaimed wastewater is used to recharge aquifers, usually
for purposes other than domestic supply. However, municipalities have
been known to recycle and treat wastewater for potable use, including
one in Texas (Bouvwer, 1992). This is exceptional. However, quite a few
cities use treated water from polluted streams for ordinary municipal
use.

Desalination

To those living near the sea and short of drinking water, the abundant
waters of the ocean must seem very attractive. However, it is quite clear
that although desalination is a well-developed and entirely feasible pro-

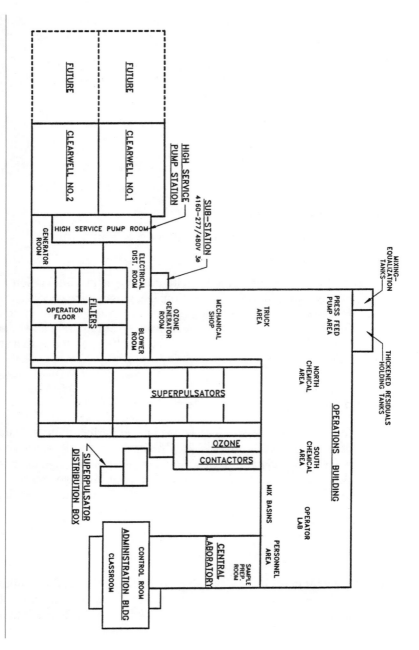

Figure 4-2. Ozonation Treatment Plant. Source: New Jersey American Water Co.

cess, it is expensive and not widely used in the United States. Worldwide there are nearly 4,000 desalination plants, about two-thirds of which are located on the oil-rich and water-poor Arabian Peninsula (Abelson, 1991). However, one really feasible site in the United States is Key West, Florida, which is so far from the mainland that a pipeline of adequate size would be prohibitively expensive. The most feasible desalination method for general use in the United States is probably reverse osmosis, which can profitably use waters that are only mildly brackish. Desalination plants may be built in conjunction with electric generation plants, profiting from the excess heat available. From time to time, desalination is given serious consideration in areas having chronic water supply problems, such as southern New Jersey, which is generally underlain with saltwater but has only a few, relatively small streams and a growing population. Cape May City recently announced that it is proceeding with construction of a small reverse-osmosis desalination plant with capacity of one mgd.

Another example of desalination in the United States is in Newport News, Virginia. In this case, water demand is growing rapidly, with great distances from any supplies of river water. However, brackish groundwater that can be treated by reverse osmosis is available. A total of seven mgd will be withdrawn from two aquifers and given two stages of reverse-osmosis treatment. If successful, the process can be extended (Suratt, et al., 1997).

Water Supply Problems

Water demand increases with the growth of population and industry. Although this growth in per capita water use has been slowed by increased prices, the widespread adoption of water conservation practices, and disposal problems will cause many communities to gradually increase their supplies of water. This puts them in competition with other areas and other uses of water. Moreover, existing supplies of water often become threatened by pollution. (See Chapter 9.) In some cases, the poor water quality must be compensated for by special methods of treatment. In some areas, groundwater supplies become permanently depleted or threatened by saltwater. Surface water supplies are periodically threatened by drought. Adjacent water supply systems need to be connected in order to meet emergency needs or temporary problems of supply, but there are institutional difficulties in such arrangements.

Although only rarely can new major projects be built, it may be possible to raise the maximum water level of an existing reservoir, or reallocate its storage. Processes are lengthy and obstacles are abundant, but it can be done.

Substantial increases in safe yield can be obtained by coordinating the operations of two or more water supply systems. The classic prototype is that of the Potomac Basin, where the total safe yield during drought was increased by over 40% without additional infrastructure. This was due to special circumstances not likely to be duplicated elsewhere, but in most cases considerable increases can be obtained. These advantages must be considerable for intersystem coordination to be adopted because system managers do not like to relinquish their independence.

Water Supply Critical Areas

The establishment of a water supply critical area is a management device to deal with areas threatened by saltwater intrusion due to excessive drawdown by wells. In some cases, the aquifer pressure has been lowered by overdrafting to 80 feet or more below sea level. If the water from adjacent aquifers of higher pressure is saline and it is drawn into the low pressure area, wells are quickly made unusable.

This figure shows the intrusion of saltwater into the water supply aquifer up to the year 1986 and the wells affected. It also shows predicted areas of saltwater pollution for the years 1990-2020, and the additional wells predicted to be affected at those times. It should be noted that there are many wells to the south (not shown on the figure) that played a major role in the intrusion of saltwater in this case.

In order to establish a water supply critical area, it is first necessary to make a thorough survey of underground water conditions, withdrawals by wells, and the area threatened. Then it must be estimated how much water is naturally recharged to the area. This will allow an estimate of the percentage by which withdrawals exceed the natural recharge. The state must now summon its courage and mandate a proportional reduction of the withdrawals from all wells in the area, together with a prohibition of further well drilling. The state must also take steps to assure that alternate supplies of water are made available. In some areas of the West, drilling of new wells has been prohibited where groundwater is

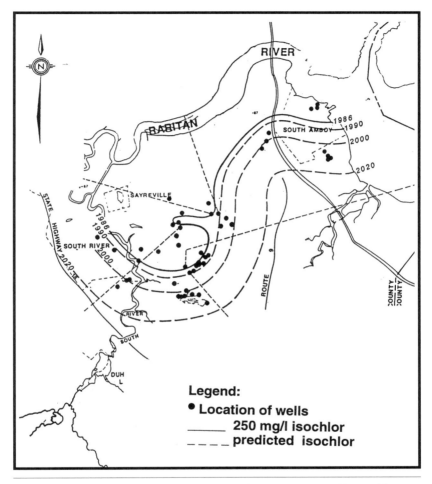

Figure 4-3. Saltwater Intrusion, near Sayreville, N.J. From *New Perspectives in Water Supply*, CRC Press, Inc.

threatened by excess withdrawals. However, in most cases, the states have not required reductions in the quantity of water previously withdrawn from existing wells.

In New Jersey, two water supply critical areas were established by the process outlined above, as shown in Figure 4-4. In both of these areas,

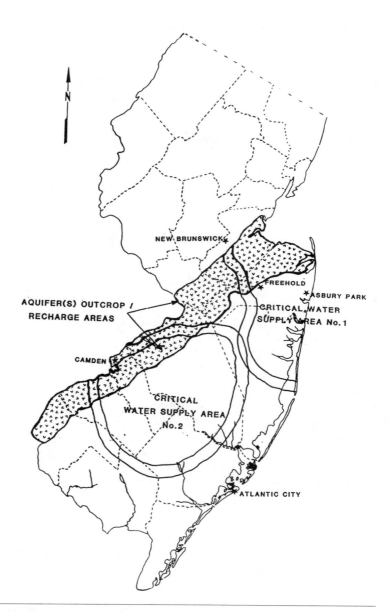

Figure 4-4. New Jersey Water Supply Critical Areas. From *New Perspectives in Water Supply*, CRC Press, Inc.

the pumping of groundwater has exceeded the amount replenished by rainfall on the recharge areas shown for a number of years. Consequently, the pressure level in the aquifer has dropped to below sea level and saltwater has intruded. In Critical Area No. 1, it has intruded from Raritan Bay (as shown in Figure 4-3) and from the sea. In Critical Area No. 2, it has intruded from "connate" saltwater that accumulated in the aquifer in ages past. For both critical areas, the state reduced authorized pumpage within the areas shown and made arrangements for alternative sources.

For Critical Area No. 1, the principal alternate supply was provided by the Manasquan Reservoir, which was built for the purpose by the state. The reservoir supplied water to three treatment facilities: the Monmouth County Improvement Authority plant, the Howell Township plant, and New Jersey-American's Jumping Brook Treatment plant. New Jersey-American subsequently acquired the Howell Township plant. The privately owned Jumping Brook plant is the principal treatment location for approximately 12 mgd of the 16 mgd drafted from the reservoir daily. New Jersey-American's investment to implement this critical area amounted to $60 million in new treatment and transmission facilities. A pipeline from the north completes service for the balance of the critical area (Whipple, 1994).

Critical Area No. 2 includes a region of southern New Jersey served by the Potomac-Raritan-Magothy Aquifer. The aquifer was being depleted because steady population growth and economic development had increased water usage. Water levels had dropped by more than 100 feet and continued dropping at the rate of two feet or more per year. The New Jersey Department of Environmental Protection designated this as Critical Area No. 2, ordered aquifer withdrawals reduced by 22% and called for development of an alternate water source on the Delaware River. The New Jersey-American Water Company agreed to plan, finance, and build the $192-million project. The finished plan consisted of a state-of-the-art water intake and purification plant on the Delaware River (described above) and more than 17 miles of 16-inch to 52-inch transmission pipeline capable of connecting three counties and 55 communities.

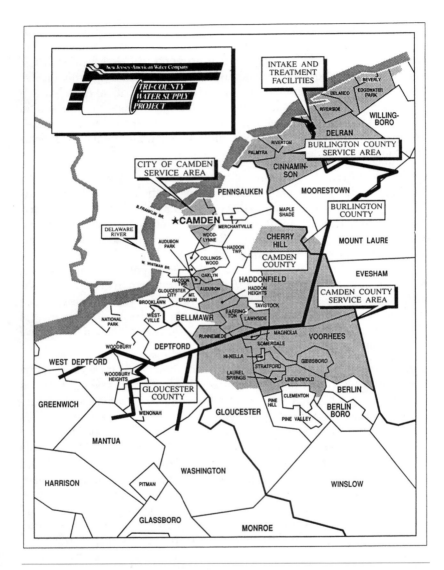

Figure 4-5. Tri-County Water Supply Project. Source: New Jersey American Water Co.

Determining the appropriate plant size involved risk: the company had to plan a system to serve not only the current and future needs for its present customers, but also to be able to accommodate those of other water systems and users who might have a future need for more water. Since all groundwater users had been ordered to cut their use by 22%, municipalities and industries either had to contract with New Jersey-American, reduce water usage, or look for alternate sources of their own. Approximately 120 permits and approvals were needed from federal, state, county, and municipal governments, and interstate commissions and regulatory agencies. The various municipalities and industries were very slow to sign contracts, so the company had to initiate a $192-million project based on faith that the state would carry through on its expressed intention to require a major reduction in groundwater withdrawals by all users. However, the planning had been soundly done, the steady deterioration of the aquifer was a fact, and the state, after considerable hesitation, moved forward with the critical area arrangement.

Water Supply Drought Management

Periodic drought is an upsetting factor in planning water supply for municipal, industrial, and private uses. Of course, other water uses may be affected by drought, including navigation, irrigation, hydroelectric power, public recreation, fish, and wildlife. Drought management for water supply is difficult because remedial measures necessarily involve both local agencies and usually also state and federal, especially if large areas are involved. However, there is no federal agency that can assume full responsibility for drought planning.

Droughts occur usually in the summer and fall, but often last a number of years. Droughts up to six or seven years are fairly frequent, and analysis of tree-ring growth records indicates that much longer droughts have occurred. Droughts are irregular in extent and may vary in severity of impact from place to place. Deficient flow in rivers during drought can often be supplemented by storage in reservoirs—either annual storage, which, for example, can capture spring rains and/or snow melt to use during periods of high demand in the summer and fall, or carryover storage, which may hold water for judicious use over a period of years. The difficulty with carryover storage is that it is unpredictable how long the drought will last and which will be the year of the most severe drought.

When a major drought occurs, the states and federal agencies must make a concerted attempt to deal with it. Local authorities must be brought into the picture. There should be consideration of measures both strategical and tactical (preventive and corrective). The plan must include providing a drought warning stage, calling for voluntary corrective measures, and declaring a drought emergency to call for more drastic mandatory measures. Drought plans may include provision for joint action between water supply systems and groundwater withdrawals (beyond the normal safe yield) for conjunctive use. There should be close coordination between state, municipal, and federal agencies, not only to formulate a plan, but to assure prompt action as the drought progresses.

Almost all drought management plans involve water conservation measures, which usually must be enforced by the municipality. They may include prohibition of car washing, use of low-flow shower heads, limitation of lawn watering to once a week, and "Plimsoll lines" on bathtubs. (Originally the Plimsoll line was on ships; on bathtubs, it is a water conservation measure indicating the depth beyond which the tub should not be filled). More severe measures may include an allocation of water per member of the household. (This requires water meters for each household, supervision of the system, and penalties.) Industrial water use may also provide important economies, including recycling of water in cooling towers (Whipple, 1994).

A striking example of overcoming water shortages through conservation is the experience of the Massachusetts Water Resource Authority, which provides water wholesale for over two million people (Kempe, 1992). The strategy included leak detection and repair, meter rehabilitation, retrofit of residential users with water-saving devices, outreach and water audits for commercial and industrial users, and public education programs. In five years of the program, total water use dropped by approximately 20%. Since the authority is a wholesaler of water, the strategy could not require conservation-related price structures, although some of the communities involved used them.

Very substantial economies can be made in irrigation use. In some cases, additional water may be purchased from other sources. It is possible to have temporary maximum pumping of wells during drought, with a subsequent reduction in withdrawals during periods of normal rainfall in order to allow the aquifer to recharge. This, however, is difficult to arrange. A more usual practice is to accept the drought situation as an indication that the regional supply of water is inadequate, and to initiate actions to develop increased sources. Another frequent course of action is to increase the withdrawal from streams, to the detriment of fish and wildlife downstream, or of other downstream users. More complex arrangements are being developed in some western states. In 1997, agreements were finalized for Arizona to store and bank, temporarily, surplus water from its authorized allotment from the Colorado River—water, which otherwise would be taken downstream by California. The water in question will be stored underground and later sold to purchasers, some of them in California (*U.S. Water News*, April 1997). Interbasin transfers are fairly common in the West but only occasionally elsewhere, as in the arrangement to divert a substantial flow from the Delaware River basin for use in New York City.

Because of the severe droughts that plagued many parts of the country in 1988, Congress authorized the U.S. Army Corps of Engineers to make a study, "The National Study of Water Management during Droughts" (Werick, 1994; Whipple and Werick, 1994; U.S. Army Corps of Engineers, 1994). The objective of the study was to develop a strategy to improve water management during drought in the United States. Case studies were conducted in four basins: the Marais des Cygnes-Osage River, the Kanawha, the Cedar and Green Rivers in Washington, and the James River. The Corps of Engineers participated in data collection and dissemination and in investigating the possible use of federal projects. The Corps also convened regional working groups, inviting states and others to participate. It provided national expertise to help solve regional problems, including computer simulations, and provided funding for the exercise. State departments of ecology and fisheries were also an integral part of the effort. The states were relied upon to ascertain the feasibility of various water conservation alternatives and the environmental effects.

The results of these studies were helpful to the states concerned. They revealed very clearly the value of the strategy of bringing together federal, state, and local interests in drought management planning. Whereas some types of water resources planning may be done by enactment of standards that are applied nationwide (e.g., point-source pollution control), and others may be accomplished with relatively little non-federal participation (e.g., traditional multiple-purpose planning), drought management planning calls for a true partnership approach. This was successfully done in the drought management studies. However, these drought studies were exceptional, made in response to a congressional request. Most states have much less adequate drought preparedness planning. It will be shown later (Chapter 11) that a partnership of this type may be useful for other kinds of planning.

Because of the continued increase in population and water demand, the need for water increases in most states, and the need for more water may be the precipitating factor in a complex multistate water problem, as it was in the ACT-ACF study. The need for more water usually results (after a drought) in a demand for more storage to produce it, or for purchase of water from more distant sources. The capital costs of providing additional water supply are borne mostly by states, municipalities, and occasionally private enterprise. As a matter of policy, concurrently with such expenditures, there should be an evaluation of both ground and surface water sources in the vicinity, and use of a system of allocation of water if one does not already exist. A water conservation program should also be adopted covering household use, industry, and irrigated agriculture, not only during an actual drought but also in advance (such as low-use toilets and showers). There is no basis for incurring large expenditures for additional water supply unless reasonable provisions are made that the water available will be economically used.

References

Abelson, P.H. "Desalination of Brackish and Marine Waters," *Science*, 251, March 15, 1991, p. 1389.

Anon. *U.S. Water News*, April, 1997. "Arizona Officials Agree to 'Bank' Unused Colorado River Water" and "Water Transfers a Major Part of San Diego's Water Plan."

Bouvwer, H. "Reuse Works." *Civil Engineering*, July 1992, p. 72.

Gaines, B. "Effluent Re-use, City of Odessa, Texas," in "Responsible Water Stewardship." Conference, January 1996, ASCE.

Kempe, M. "Solving MWRA's Supply Issues through Conservation," in *Water Resources Planning and Management*, ASCE 1992.

New York Times, May 25, 1997. "To Support Growth Albuquerque Will Shift Source of Water."

Revkin, A.C. "New York City Sued by U.S. on Water Filtration Plant," *New York Times*, p. B4, April 25, 1997.

Solley, W., Merk, C. and Pierce, R. "Estimated Water Use in the United States, 1985," U.S. Geological Survey, Washington, DC, 1988.

Suratt, W.B., St. John, G.A., and Harris, R.E. "Newport News: Brackish Water Gets Fresh," *Civil Engineering*, July, 1997, p. 55.

U.S. Army Corps of Engineers, "National Study of Water Management during Drought," IWR Report, 94-NDS-12, September 1994.

U.S. Water Resources Council, "Economic and Environmental Principles and Guidelines," March 10, 1983.

Werick, W.J., "National Study of Water Management during Drought," U.S. Army Corps of Engineers, IWR Report 94-NDS-12, September 1994.

Whipple, W., Jr. and Werick, W.J. "Drought and Water-Supply Management Roles and Responsibilities." Discussion: *Jour. Water Resources Planning and Management* 120, 6, 1003. November-December 1994.

Whipple, W., Jr. *New Perspectives in Water Supply*, CRC Press, 1994.

Chapter 5

FLOODS AND FLOOD CONTROL

Major Floods of the Past

No one who has seen a major river at maximum flood stage can fail to be deeply impressed. There was an awesome flood in 1937 on the Ohio River, with ten miles of muddy turbulent water across the river between Cairo, Illinois, and the opposite bank in Kentucky, with whole trees, dead cattle, pieces of houses, and endless trash floating by. The river was up to three feet higher than the levees protecting the town, with hastily built walls of planks and sandbags on top of the levees precariously holding the waters back. When waves sprang up on the river, flood fighters had to keep a strict watch. In 1997, there was another flood on the Ohio, of lesser magnitude but still impressive.

While in high school, the author witnessed floods on the lower Mississippi, one mile wide between high levees. On one occasion he swam the flooding river, accompanied by two other boys in a row boat. On another occasion, the river started to overflow across the top of the levee, and a fast runner was sent to spread the alarm. There was good reason. The river was more than 20 feet higher than the surrounding flat countryside.

Such matters are now considered ancient history. We have flood control plans for the Mississippi, the Ohio, the Missouri, the Columbia, and many other major rivers. Even though extremes of flood-producing hydrologic events may still occur, the reservoirs, dikes, levees, and channel improvements will reduce the flood impacts. However, the flood plans on the major rivers do not provide complete protection, as the damaging Mississippi/Missouri flood of 1993 convincingly demonstrated. There are also many rivers of smaller sizes where flood control plans were never completed. Floods may occur as a result of rainfall, snow melt, dam breaks, geologic changes, or ice jams, in spite of partial con-

trol by dams, channel improvements, levees, and floodplain management.

Figure 5-1 shows the North Fork Kentucky River in flood in 1963. It is
an aerial view of several miles of river, showing the concentration of
community development along the river and the virtually complete inundation of the floodplain, with all its buildings. Figure 5-2 shows flood
damage in Wilkes-Barre, Pennsylvania, from the Susquehanna River,
1972. It is apparent how seriously flooding of this intensity can affect a
town. Figure 5-3 shows the virtual destruction of a home by the Eel
River, California. Figure 5-4 shows a flood wave hitting Putnam, Connecticut, in 1955. (Although floods, especially on large rivers, usually
rise very slowly, there are sometimes flood waves, usually caused by
the breaching of a dam upstream, or a cloudburst flood. Such a wave
could be very destructive.)

Figure 5-1. North Fork Kentucky River in Flood. Courtesy of
American Red Cross.

Figure 5-2. Flood Damage in Wilkes-Barre, Pennsylvania. Courtesy of American Red Cross.

Floodplain Management

A study of the history of flood control improvement shows that, in spite of the completion of new projects, the total of flood damages does not decrease proportionately. This occurs because as reservoir flood control makes the higher portions of the floodplain less likely to flood, that land is taken to build houses, or is used for business purposes. This is usually less flagrant than what happened some years ago in the vicinity of Branson, Missouri, where the lower portions of the town had been frequently damaged by flood. A flood control reservoir, the Table Rock project, had been authorized and was under construction upstream when an officer of the Corps of Engineers visited the area and found several hundred house foundations under construction in a part of the floodplain below the town, in effect creating at a lower level the flood hazard the project had been built to remedy. That construction was summarily interrupted.

Figure 5-3. Flood Damage from the Eel River. Courtesy of American Red Cross.

It was Gilbert White (White, G., *et al.*) who first brought to national attention the fact that in spite of the nation's heavy investment in flood control structures, flood damages appeared to be actually increasing. Government action followed in the interest of a more balanced approach to reducing damages in floodplains. Action was encouraged by the Water Resources Council (now nonexistent). The Flood Disaster Protection Act of 1973 provided incentives for states to establish zoning controls for floodplain management, particularly in the floodways, but also in some states' flood fringe areas. Some consideration was given to flood-proofing vulnerable structures and to acquiring land in the most exposed areas for purposes of recreation or wildlife, but the main action was floodplain zoning.

Figure 5-4. Flood Wave Hitting Putnam, Connecticut. Courtesy of American Red Cross.

In 1993, the disastrous Missouri-Mississippi flood resulted in convening an "Interagency Floodplain Management Review Committee" (IFMRC). Its report considered existing floodplain management strategies and recommended major changes (IFMRC, 1994). It recommended that environmental quality and national economic development be established as twin objectives of water resource development. The linkage between floodplain protection, habitat protection, and disaster programs should be strengthened to the advantage of both goals, and needed lands should be acquired. Collaboration between federal, state, and local agencies should be sought. These recommendations appear cogent today, but have not been fully carried out. The report also recommended improvements in the flood insurance program, which have since been implemented (Cunniff and Galloway, 1995).

Section 73 of the Water Resources Development Act expressed the will of Congress to equalize federal participation in structural and in non-structural approaches to controlling floods, including, besides zoning, flood proofing and acquisition of land (Platt, 1980). Although this impressive policy has not been implemented to any great extent, it remains true that, with increasing difficulty in implementing reservoir flood control, floodplain management is getting further emphasis. The availability of flood insurance at reasonable rates helps obtain agreement on floodplain zoning.

This action was all constructive; it slowed, although it did not stop, the development in floodplains (Tucker, 1978). Meanwhile, four aspects combined to maintain or increase the flooding. First, the construction of levees and floodwalls automatically reduces the valley storage available (in the area protected) and correspondingly increases flooding elsewhere. Secondly, channel improvements speed the water downstream to increase flooding below. Thirdly, the construction of homes, businesses, and transportation facilities increases runoff because of the high proportion of impervious area. Even ordinary homes have considerable paved surfaces and roofs, and any small hollows, swamps, or meandering brooks are usually filled in or smoothed out in the grading process. The fourth aspect, while important, is not so obvious. If small detention basins are built to detain runoff from only a few acres of ground, they are designed to protect against the relatively frequent small storms with a concentration time of perhaps half an hour. Floods on even moderate-sized rivers may have a concentration time of perhaps 24 hours. The steady rain for 24 hours that produces a flood for the larger stream will be allowed to pass quickly through the outlets of conventionally designed small detention basins. Larger rivers may have concentration times of weeks or even months, and small detention basins covering even the entire drainage basin would have virtually no effect upon the flood flows of such a river. Accordingly, construction of conventional stormwater management detention dams has virtually no effect in reducing floods on the larger rivers. (Detention dams now being built with provisions for water quality are more effective.)

Levees, channel improvements, and floodways must be incorporated into the plan with great care, in view of their possible adverse side-effects. Although efforts have been made to decentralize-flood protection efforts, jurisdictional problems are bound to arise, since boundaries are more apt to follow streams and rivers than watershed limits.

Special Districts

One of the best ways to plan and administer flood protection is the establishment of special districts for the purpose. The outstanding example is the Denver Urban Drainage and Flood Control District in Colorado. It combines functions of flood damage protection and drainage and also helps municipalities to develop requirements for quality of runoff. It is large enough to have a competent staff that is fully informed on all the complex issues involved. It does outstanding work in improving streams for flood control while building up environmental values and improving areas for recreation, especially walking and biking along the streams. However, special districts are exceptional.

Benefit Evaluation

According to the "Principles and Guidelines" (U.S. Water Resources Council, 1983), flood control benefits are evaluated differently for agricultural flooding and for "urban flood damage." (These are economic benefits. Environmental benefits should be considered separately.) The instructions are very detailed, but the main points are as follows.

In principle, damage reduction benefits to agriculture are the increase in net income due to the flood control plan such as increased crop yields and decreased production costs. This seems appropriate.

As regards "urban flood damage," benefits are primarily the reduction in actual or potential damages associated with land use. However, if an activity is added to the floodplain because of a flood control plan, the benefit is different in aggregate net income with and without the plan. Flood damages include physical damages, income losses, and emergency costs. Prices used for evaluation should be those prevailing during the period of analysis. For water-based recreation the agency determines unit values.

Both agricultural and urban damage control benefits are evaluated as National Economic Benefit (NED). Presumably, the destruction of a fruit orchard in California would be excluded from damages if it were accompanied by increased marketing of the same fruit from Florida. However, this relationship is not specifically covered, and is difficult to establish.

Benefit/cost ratios are not the sole criterion used to determine the extent of federal investment in flood control. Intangible benefits, including prevention of loss of life, can also be considered. The unwritten (but very effective) policy has been to provide for control of at least a 100-year-frequency flood, for each project, even if economic analysis might show protection against the more numerous small floods to have a higher B/C ratio. Where there is as yet no flood control plan, the occurrence of any really disastrous flood has the effect of changing agency viewpoints very quickly, as was the case in both the Ohio River and the Columbia River.

Flood Control Storage

For all but the small local areas, flood control storage must be carefully and deliberately planned. In addition to standard design, cost, and benefit studies, coordination may be required with other project purposes, such as water supply, as has customarily been done in traditional multiple-purpose studies. However, any major flood control study (and particularly the estimate of benefits) must also be accompanied by analysis of alternative non-structural methods, particularly land acquisition, flood proofing, and floodplain zoning. Also, possibilities of favorable environmental features should be considered.

It is usually desirable to plan protection of major flood hazard areas against all normal floods. The standard measure of protection is the 100-year flood. Occasionally, protection will be given for occurrence of the Standard Project Flood, which is larger. Of course, spillways for earth dams are designed to handle an even greater discharge, the Spillway Design Flood, since failure of the spillway would destroy the dam, with disastrous consequences downstream. (However, the very small dams are built with much reduced spillway design floods.) Figure 5-5 illustrates the dependence of the homes below a dam on the soundness of the dam/spillway construction.

The management of flood control storage during a flood may be a complex matter. Reservoirs are designed for a given amount of flood control storage that is only occasionally needed, but the shores of the reservoir are much used by wildlife and also for moorings for boats and other recreation facilities. Accordingly, once the storage is filled, there are demands to lower the level as fast as practicable. Moreover, there is

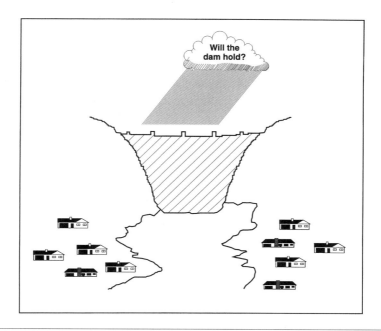

Figure 5-5. Necessity for a Spillway.

always the possibility of another storm occurring when the reservoir is already full and incapable of further storage. On the other hand, releasing the stored flood waters prolongs the inundation of floodplains downstream, which occupants may be anxious to get the use of. Altogether, the use of such storage requires good judgment and the reliable predictions of weather.

Runoff Pollution

The handling of stormwater runoff was greatly complicated by the discovery (not long after the passage of the Clean Water Act) that the greater part of the pollution in streams generally comes not from the point wastes (towards which that act was primarily directed) but from "nonpoint sources." When this fact was revealed to an official of one state by a university researcher, instead of congratulating the university for innovation and successful research, he burst out, "My God! I hope this doesn't come to be known". (He was concerned with the fate of a state bond issue that had been publicized on the implicit assumption that the known point sources were all that had to be considered.) Since that time, great progress has been made, and the ways of dealing with pollution in run-off are now fairly well understood, as indicated in Chapters 8 and 9, although actual progress has been very irregular.

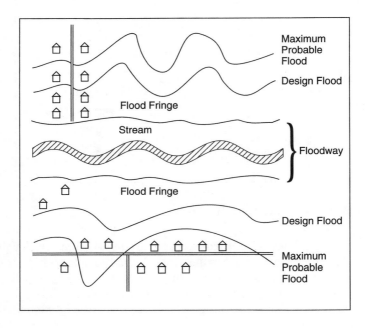

Figure 5-6. Diagram of Floodplain Zoning.

Environmental Use of Floodplain Zoning

Historically, floodplain zoning has been established with the idea of pre-venting the building of homes and businesses in floodplains in order to reduce flood damage.   See Figure 5-6.

Floodplain zoning, as shown in this figure, provides a frequently flooded floodway within which no construction is allowed except for indispensable structures, and a flood fringe area in which construction is limited. Any building permitted may have foundations raised out of reach of a 100-year flood or be otherwise flood-proofed. In some states such as New Jersey, the amount of fill that can be added by new construction in the flood fringe area is limited. The usual result, in other than urban areas, is that the floodway and most flood fringe areas remain naturally wooded or wetlands, and good refuges for birds, insect life, fish, and native plants. Although forestry and grazing lands can result in destroying a lot of woods, this does not have to be the case. With a few other restrictions, properly designed floodplain zoning can provide environmentally important strip corridors along most sizable streams. In partial recognition of this relationship, the Association of State Flood Plain Managers now encourages use of wetlands managers.

The Vicksburg District Corps of Engineers has found that numerous water control structures, confined disposal facilities, and borrow pits being built for flood control can be adapted to produce environmental benefits by impounding water. The intent is to compensate for habitat losses that would result from other features of the project (Fischenich, *et al.*, 1993).

Summary

In summary, flood control on major streams should be developed by engineering analysis on a multiple-purpose watershed basis. To be complete, the flood control aspects must include floodplain management. If protection of the smaller rivers and streams is to be included, systems of small impoundments are needed. The plan for impoundments must be based upon systems of land use control, but should include provisions for controlling quality of runoff. (See Chapter 9.) It would be extremely difficult for all of this to be included in a single study or implemented through a single program. In later chapters, flood control is considered first as an aspect of standard multiple-purpose developments and secondly, it is used as part of a program including land use and nonpoint source control, aimed largely at water quality.

Moreover, there remains the major environmental benefit that may be gained from floodplain zoning and environmental use of flood control features. Altogether, flood control, originally considered to be a rather simple matter, now becomes very complex, with the inclusion of important environmental aspects.

References

Cunniff, S.E. and Galloway, G.E. "Sharing the Challenge of Effective Flood Plain Management," in "Integrated Water Resources Planning for the 21st Century," ASCE 1995.

Fischenich, J.C., Dardeau, E.A., and Parrish, K.D. "Use of Flood Control Structures for Environmental Mitigation," in *Water Management in the '90s*, ASCE, 1993.

IFMRC 1994. Interagency Floodplain Management Review Committee. "Sharing the Challenge: Floodplain Management into the 21st Century," Washington, D.C.

Platt, R. "Intergovernmental Management of Floodplains," University of Colorado, 1980.

Tucker, L.S. "Flood Insurance and Floodplain Zoning," in *Water Problems of Urbanizing Areas*, ASCE, 1978.

U.S. Water Resources Council, "Economic and Environmental Principles and Guidelines for Water and Related Land Resources Implementation Studies," Mar. 10, 1983.

White, G., et al. "A Unified National Program for Managing Flood Losses," HD 465, 89th Congress, 2d Session, Report of the National Task Force for Flood Control.

Chapter 6

NAVIGATION, HYDROELECTRIC POWER, AND IRRIGATION

Navigation, General

Navigation was the first water use to attract national attention. It is hard for us to realize how essential water transportation was to our country before the development of our present systems of railroads and highways. In New Jersey, the 65-mile-long canal, the Delaware and Raritan Canal, links the Delaware River to the Raritan River, near New Brunswick. It is used for water supply and also as a strip corridor park. However, it was built entirely for navigation, to move coal barges drawn along a towpath by mules, from the coal-rich regions of Pennsylvania to the metropolitan New York area.

As our country spread to the West, explorers, trappers, and adventurers used the rivers as means of transportation, and many cargoes from the central part of the country were shipped down the Mississippi River to New Orleans for export. Consequently, much of the early thinking about our rivers envisaged them as means of transportation to be used together with ships at sea and in the Great Lakes. Even after full development of our railroad and highway systems, our rivers remained the most economical means of transportation for bulk cargoes. On the larger rivers this is done by large barges, assembled in tows of perhaps fifteen barges and pushed by a single tug boat.

Sometimes navigation is on the open river (such as the Missouri below Yankton, South Dakota, and the lower Mississippi). Other rivers have locks and dams, such as the Ohio, the upper Mississippi, and the Columbia above Bonneville Dam. The locks are preferably of a size to accommodate a tow of barges without the necessity of disassembling it and passing the separated barges. The depth of the navigation channel, including the locks, is a key factor governing the economy of river transport.

The locks on the Ohio, which, when built, were considered to be a tremendous engineering achievement, are now being painfully (and expensively) reconstructed, with locks 1,200 feet long and a great reduction in the number of locks and dams.

Current Navigation Programs

Further major extensions of inland navigation in the United States are unlikely to be undertaken. More usual problems may involve proposals for deepening and improving channels in our ports and estuaries for seagoing traffic, reconstruction of obsolete locks, routine channel maintenance by dredging, and maintenance of river flows adequate for navigation. In times of drought, the maintenance of river flows adequate for navigation is competitive with other uses of water. On waterways with a high traffic in petroleum products, such as the Delaware Estuary and the lower Mississippi, water pollution due to spills and leakage is an important problem.

Benefit Values of Navigation

The benefit of water transportation is evaluated as the difference between the costs of using the waterway and those of using alternative means of transportation (U.S. Water Resources Council, 1983). However, by law, comparison of prevailing rates may be used to represent costs for different modes of transportation. On an economic basis, the officially sanctioned comparison favors the waterways, as rail rates must include the cost of amortizing and maintaining the investment in rail infrastructure (with some reduction for land-grant rail lines), while water rates include no part of the construction or maintenance costs of the waterway. The waterways' benefits are usually computed on the basis of commercial cargo handling, but there is also a great deal of recreational boating on these waterways.

Hydroelectric Power, General

Hydroelectric power is one of the more recent forms of government water resource development. As a means of generating energy, it is very clean, without the problems of waste heat and polluted exhaust that accompany thermal generation from coal, or the problems of dangerous waste and possible accidents from nuclear plants. Other alternative modes of generation are tidal, geothermal, and solar power, but they are exceptional.

Hydroelectric Power Operations

In operation, hydroelectric power plants are very flexible. Whereas nuclear power plants must operate in a steady state, and large steam plants take some hours to start or to stop production, a hydroelectric turbine backed by a reservoir can readily be varied in operation in a matter of minutes, without much loss in efficiency. This is very important for system operation, as the demand for power characteristically fluctuates widely during the day and declines to very low levels at night. Also, there may be outages due to lightning or equipment failure.

To provide for flexibility in a large system, it is possible to have a special type of hydroelectric power plant, the pumped storage plant, strategically located near a major load center. This plant operates between a lower- and an upper-level water impoundment. Water drawn from the upper impoundment generates electricity as it is discharged into the lower impoundment (or river). Alternatively, the pumped storage plant can use surplus electricity to pump water back into the upper reservoir, in effect providing a reservoir for electric energy.

Controversies Related to Hydroelectric Power

Although hydroelectric plants have great inherent advantages, they have major environmental disadvantages as regards fish. The rapid variation in the velocity of flow downstream of the dam to meet varying power requirements is harmful to certain fish. However, problems with anadromous fish are more important, particularly salmon. Power dams are usually located with deep impoundments of water (in order to gain high efficiency), and they usually form a barrier to passage of anadromous fish.

These problems with fish are raised in connection with about 1,600 nonfederal power dams. Environmentalists have been pressuring the

Federal Energy Regulatory Commission to force the owners to operate these dams in a more environmentally friendly fashion as a condition of renewing their licenses. The staff of the commission, after intensive consideration, has recommended the removal of the Edwards Dam on Maine's Kennebec River, in order to obviate its adverse effects upon anadromous fish (Malakoff, 1997).

For certain projects the adverse effects upon salmon was foreseen long ago, and the dams on the Columbia and many other rivers have fish ladders, which are quite efficient in allowing movement of adult fish migrating upstream in order to spawn. However, the problems of moving the immature "fingerlings" downstream after spawning have proved much more difficult to handle. Not only may the fish be damaged if they pass through high-head turbines, the inevitable delay in moving downstream through a large reservoir exposes them to lack of food and to predation from other fish. Various ways to deal with this problem have been devised. Environmentalists have urged a basic restructuring of the Columbia and Snake River system in ways that would benefit the fish but greatly reduce the production of hydroelectric power. This would include lowering Columbia River reservoirs and the four dams on the lower Snake River to the lowest level at which navigation locks and the irrigation pumps could still operate. In addition, 1.2 million acre-feet of storage from the Dworshak, Brownlee, and Upper Snake River reservoirs would be released in order to speed up the river. On the Columbia River, up to 3 million acre-feet of water would be released, in addition to the 3.45 million acre-feet released under existing arrangements (Ruff and Fazio, 1993).

A more modest counterproposal, which was put into effect in 1994, is the McNary Juvenile Fish Facility at McNary Dam (ASCE, 1997). This includes a fish-collection channel and associated dewatering/control facilities, a fish transport pipeline, and a bypass system, including fish holding and loading facilities. This system moves juvenile "fingerling" salmon onto barges and trucks for transport past the three remaining dams and reservoirs of the lower Columbia River.

It is interesting to note that some analyses of the loss of salmon in the Northwest list four factors other than dams as responsible for the loss: habitat loss and degradation, over-exploitation in sport and commercial fisheries, variable ocean conditions, and effects of hatchery practices. Among the habitat conditions cited as responsible are large fires and intense winter rainstorms. The salmon population has some natural re-silience to overcome such disturbances, in an example of dynamic equi-librium. In order to profit from this resilience, timber harvesting must be designed so as to correspond as nearly as possible to natural changes (Reeves *et al.*, 1997). In Oregon the governor recently proposed spend-ing $30 million to clean up streams and restore fish runs, as an alterna-tive to federal action under the Endangered Species Act (De Sena, 1997). It is apparent that this problem is far from solved. In view of the national prominence of the dam/salmon problem of the Columbia River, further progress is anticipated on systems to save the endangered species of salmon without crippling the immense potential for hydroelectric power production.

Somewhat different fish problems are encountered on the Colorado River. Below the immense Glen Canyon Dam, the river, which before the dam was scoured by large annual floods, now flows more irregu-larly, commensurate with the need for electric power. Vegetation and birds have flourished in the new regime. Yet some environmentalists, including the Sierra Club, wanted to destroy the dam, since the cold, fluctuating flows, while favorable to rainbow trout, are harmful to some other fish, including one of the endangered species. The Secretary of the Interior responded to the criticism by mandating an upper limit of releases from the dam of 25,000 cubic feet per second (cfs), and a con-firming study to find the optimum situation. There are also many prob-lems related to the movement of sediment in rivers below the large dams on the Colorado (Collier, 1996). High flows are useful to scour the chan-nel bed and expose gravel that is necessary for egg attachment, and to protect small fish in backwaters. Releases from the dam during the year are used to promote native fish habitat as much as possible, and may be used to rebuild eroded beaches. In 1996, a large artificial release of 45,000 cfs was made to test these effects.

Benefit Values of Hydroelectric Power

The "Principles and Guidelines" (U.S. Water Resources Council, 1983) gives somewhat inconsistent advice as to the basis for evaluating the benefits from energy produced by hydroelectric power plants. In Sec-tion 2.5.2, it is stated that the conceptual basis is society's willingness to pay, and if that market price corresponds to marginal production cost, it may be used as a measure of benefits. However, the same paragraph explains that because of different financial conditions, insurance, and taxes, a special procedure is required to give the real resource cost of the most likely federal alternative. The procedure specified (Section 2.5.8) requires calculation of the costs of the alternative energy source, using federal discount rates, with no taxes or insurance charged to the alternative. This procedure will give "resource costs" much lower than either marginal or actual prices charged. This is a strict standard, with no inherent advantage for the federal project.

The idea of comparing federal projects to private ones for a tax-equivalent basis arose in the early 1950s and caused considerable discussion and some defiance within the government. A detailed account of this matter is available, including action by government applying the tax-equivalent comparison (Whipple, 1962). However, also in 1962 the Water Resources Council issued its "Policies, Standards and Procedures…," which referred to "willingness to pay" as the measure of benefit, with no reference to a tax-equivalent basis for comparing costs or prices. So that, after an early interest in, and some application of, the idea of tax-equivalent comparisons, government policy firmly turned against it in 1962, but renewed it in 1983 with an addition of interest- and insurance-equivalent comparisons. However, there is some doubt how far these 1983 principles have actually been applied. The Corps of Engineers (which handles most federal hydroelectric projects) issued a procedures manual in 1991 (U.S. Army Corps of Engineers, 1991) that simply fails to deal with the question, referring instead to the general principle of willingness to pay as the measure of benefit. Note that both the 1962 "Standards and Procedures" and the 1983 "Principles and Guidelines" had referred to willingness to pay.

The center of interest in water resources matters has shifted away from major dam projects, so this anomaly has not attracted much attention. The abolition of the Water Resources Council has left such ambiguities with no agency responsible for clarifying them.

Western Irrigation and Subsidies

Starting in the 19th century, irrigation developed very naturally in the low-rainfall areas in the West. In most cases, water diverted from streams was involved. Large diversions and large irrigated areas were used for economical production. Frequently, substantial diversion works or dams were required. It quickly became apparent that government action was essential. The western states arranged to grant water rights for each development, and the U.S. Bureau of Reclamation was created to build and operate the larger projects on terms very generous to the landowners. The Bureau of Reclamation quickly became one of the most popular and influential branches of the federal government in the 17 western states.

Since the larger diversions required building of dams, which could very profitably produce hydroelectric power, Congress soon allowed the use of the revenues from the power to subsidize the irrigation developments. Grand Coulee is the prime example. It is a splendid power project. Extra power is available to pump the water hundreds of feet above the reservoir to irrigate a large area of uplands that otherwise could be used only as cattle range. However, other projects did not combine irrigation and power production so naturally. There was one large projected irrigation project, which, however, was too expensive to be built without a subsidy of some kind. Someone had a brilliant idea—the irrigation was to be combined in a single project with *oil wells* that were to be drilled on some government land nearby! This was too radical an idea for Congress to support. However, the next year a project under a new name was recommended; it combined the same irrigation and diversion that had originally been proposed, and a hydroelectric power plant some miles upstream, instead of oil wells. This was duly authorized and built.

Of course, from a strictly economic viewpoint, it was improper to justify an otherwise infeasible irrigation project by using surplus revenues from another separate project, whether oil wells or hydroelectric power. However, Congress, in effect, decided that the value of irrigation to the western states was so great in developing a balanced economy that it was worth a federal subsidy from hydroelectric power. But not from oil wells. That was too obvious.

Irrigation Elsewhere

In the southeastern states, a considerable growth of irrigation has developed on an entirely different basis. These states have no Bureau of Reclamation to organize projects. Some of the states have water allocation systems and others still have riparian rights. Much of the irrigation comes from groundwater. Instead of irrigation ditches, spray irrigation is usually utilized. Fruit and high-value vegetables are more frequently irrigated than the basic crops. Rice irrigation in the Mississippi Delta is an exception. Instead of being federal projects, most irrigation developments are privately owned.

Pollution and Groundwater Depletion from Irrigation

In the West, the large irrigation projects often result in difficulties with water quality downstream. The return flow from the multitude of irrigation ditches can, of course, carry pesticides and fertilizers, but also a lot of salts from the soils, particularly in the Southwest. In the East, saline soils are not so common, and sprinkler irrigation does not generate as much return flow. However, the heavy pumping from groundwater not only may reduce water available for other purposes, but in areas near the coast it may cause intrusion of saline water. Also, what is not so easily visualized is that the pumping of groundwater results in reduction of the low flows of streams. Heavy withdrawals of water from shallow aquifers can reduce the flows in adjacent streams and affect the viability of water allocation on those streams. Conversely, in some situations, where flow in streams recharges adjacent aquifers, over-allocation of stream flows may result during drought in shortages of water in nearby wells. Systems of water allocation must consider groundwater and surface water as a whole, particularly in regards to drought conditions.

Valuation of Irrigation Benefits

The *Principles and Guidelines* (U.S. Water Resources Council, 1983) has established benefit evaluation for agriculture without any hint of the special consideration given to irrigation in certain Bureau of Reclamation projects. The discussion is very detailed, but main points are as follows. In concept, the NED benefits are the value of increases in the agricultural output of the nation, and the cost savings of maintaining a given level of agricultural output (which had been threatened). The benefits from production of crops other than basic crops may be offset by a decrease in production elsewhere. If not so limited, the net value of the increased production or the increase in net income may be the measure of benefit. In effect, instead of irrigation being given subsidies and special treatment, as it was so many years in the West, irrigation is now to be treated as a subset of agriculture, as one of many economic activities. New federal irrigation projects are now very unusual, but benefits to irrigation can still be provided by multiple-purpose projects or extensions of existing irrigation.

References

ASCE, "McNary Juvenile Fish Facility, McNary Lock and Dam," ASCE News, Mar. 1997.

Collier, M., Webb, R.H., and Schmidt, J.C. "Dams and Rivers: A Primer on the Downstream Effects of Dams," U.S. Geological Survey Circular 1126, 1996.

De Sena, M. "Science and Economics of Salmon Recovery Debated" in *U.S. Water News*, p. 18, April 1997.

Long, M.E. "The Grand Canyon," *National Geographic*, July 1997.

Malalakoff, D.A. "Agency Says Dam Should Come Down," *Science*, Vol. 277, p. 762, 8 Aug. 1997.

Reeves, G.H., Brenda, L.E., Burnett, K.M., Bisson, P.A., and Sedell, R. "A Disturbance-based Ecosystem Approach to Maintaining and Restoring Freshwater Habitats of Evolutionarily Significant Units of Anadromous Salmonids in the Pacific Northwest," in "What is Watershed Stability," University of California, Water Resources Center Report No. 92, April 1997.

Ruff, J. and Fazio, J. "Columbia River Basin Fish and Wildlife Program Strategy for Salmon," in *Water Management in the '90s*, ASCE, 1993.

U.S. Army Corps of Engineers, "National Economic Development Procedures Manual," Oct. 1991.

U.S. Water Resources Council, "Economic and Environmental Principles and Guidelines for Water and Related Land Resources Implementation Studies," Mar. 10, 1983, section 2.5.2 and 2.5.8.

U.S. Water Resources Council, "Policies, Standards and Procedures in the Formulation, Evaluation and Review of Plans for Use and Development of Water and Related Land Resources," May 29, 1962, p. 10.

U.S. Water Resources Council, "Economic and Environmental Principles and Guidelines for Water and Related Land Resources Implementation Studies," Mar. 10, 1983.

Whipple, W., Jr. "Economic Feasibility of Federal Power Projects," *Land Economics Vol. XXVIII*, 3, 219-230, Aug. 1962.

Chapter 7

MULTIPLE-PURPOSE (COMPREHENSIVE) PLANNING SYSTEM

Introduction

During the period of rapid construction of water resources in the United States (1932-72), all federal project planning was at least nominally guided by comprehensive plans and amendments to them. These were first the "308 plans" and later Level B studies. These plans were really not fully comprehensive. They concentrated on the functions of the construction agencies (COE, USBR, and TVA) and paid little attention to water pollution, agriculture, environmental matters, or stormwater runoff, except to try to avoid such problems, and deal with them only to the minimum extent. Although they represented a great broadening of viewpoint, compared to the single-purpose plans that had preceded them, it is a little misleading to call them comprehensive today.

Our existing national water resource systems are mostly excellent, but we should not stand in awe of the early processes used to plan them. The Tennessee Valley Authority (TVA) was probably the one where early multiple-purpose planning was most successful. When the TVA was created in the excitement of Franklin Roosevelt's presidency, it got a flying start and made a reputation by simply building its program based on the Corps of Engineers' 308 report. The TVA was created with high aspirations to assist the economic and social development of a generally poor region. However, it also took over functions handled by various federal and state agencies and private utilities. (In later years, when attempts were made to replicate the TVA in various other regions, the idea was resisted and no others were authorized.)

The famous Pick-Sloan plan for the Missouri River was a case of a highly irregular planning process. Although given national celebrity as an example of interagency planning, it was really no such thing. It was simply a practical compromise in which the Corps of Engineers built the system of huge reservoirs on the main stem of the Missouri and a lot of flood control projects, mostly in the lower basin, and the Bureau of Reclamation built hydroelectric power and irrigation projects further upstream. Each service identified plenty of projects to work on for years to come, and it was easy for Sloan, who was a brilliant engineer, and Pick, who was a shrewd politician, to see that the Missouri Basin could accommodate large programs for both of them, to mutual advantage.

In the Ohio Basin, where the Corps of Engineers had done its very first water resources work, years of planning had led to the decision that what the Ohio needed was a series of navigation dams, with levees and flood walls to protect the cities and industrial centers. However, there came the monstrous flood of 1937. Within a few months the Corps of Engineers had sent a revised plan to Congress including a major system of flood control reservoirs, which has since been built. (However, even today, with higher dams on the main stem, there is no hydroelectric power generated, despite the vast potential.)

Something rather similar happened in the Columbia River Basin. In 1948, a complete revision of the comprehensive report had been underway for four years, at a cost of over $4 million (an immense sum for a report in those days). The plan was to extend navigation up the Columbia and Snake rivers, with huge projects for hydroelectric power generation, but no flood control storage for the main-stem floods. The complete plan was due for submission to the Chief of Engineer's office on 1 November, but a one year extension had been requested. A major flood came in June, 1948, with damages far worse than expected: $100 million in damage and 22 deaths. While the flood was still in progress, General Wheeler, the Chief of Engineers, came for a visit, flying in at midnight. In the morning there was a very short conference. After a few preliminaries, he said, "You have requested a year's extension of time to submit the basin report, but with all the national interest, there can be no delay." He paused to let the thought sink in.

"Now does the plan as it stands include a flood control plan?"

"No, sir, it does not."

"But when it is submitted, it will, will it not?"

The district and division engineers were silent and glanced at a staff officer. He answered, "Yes, sir, when it is submitted it will."

Wheeler said, "I see that my confidence in the district and division engineers has not been misplaced. Now gentlemen, I think we have time for a cup of coffee before the press arrives".

That day the Chief of Engineers made headlines with the announcement of the new flood control plan to be submitted on 1 November.

What followed was much more akin to logistical planning in wartime than to the normal deliberative process of water resources planning. There was complete delegation of authority. The provisions for flood control were incorporated without reducing production of prime power. However, it would have been much better if the damage of a major flood had been forecast earlier, and if, for example, the necessary coordination between flood control and production of power had been worked out in more detail. Not all of the quick decisions of the Corps of Engineers have been as successful as this one. Careful planning is essential to success, even though, at times, some of the processes may have to be abridged.

The Objectives and Concept

Under normal procedures, the planner first evaluates the problems and opportunities facing the basin. Usual problems are floods, droughts, and environmental damage. Possible opportunities may be hydroelectric power, flood control, irrigation, water supply, and public recreation. The floods and droughts are the most difficult to evaluate because of the unpredictable nature of extreme events and their tremendous impact when they do occur. Sometimes public or political expectations must influence the decision. In most cases, it must be recognized that there is a conflict between objectives: one objective cannot be fully developed without limiting the achievement of another objective, or causing a related problem.

Potential Projects and Screening Alternatives

Once the objectives are clear, there must be a selection of potential means for achieving them. After preliminary reconnaissance, there must be field surveys and subsurface explorations for the most promising alternatives. Cost estimates must be made, and water requirements for each potential project outlined. A preliminary choice of the most favorable potentialities can now be made. Environmental and other intangible potential problems must also be evaluated.

Selection Process

Preparations are now completed for the most complex part of the planning process, the evaluation of alternatives. Modern computer technology greatly facilitates this task. The end result is a tabulation of the more favorable potential projects, as related to drought water budgets and flood period routing, as appropriate. Benefit/cost analysis will give a preferred alternative as between reserving more storage for water supply or for flood control, for example. For each likely combination of projects, their drought water budgets will be associated with corresponding environmental and other intangible consequences of each alternative that also must be considered.

In the traditional multiple-purpose planning procedure, the above information was only very imperfectly arrived at since computations had to be made without aid of a computer.

There remained the stage of obtaining public input and coordination with other interested agencies. In modern times this stage of planning has caused increasing problems. Difficulty arises, particularly with environmental problems, but also with conflicts of interest between states, or between different public viewpoints. Such difficulties can be minimized by advance coordination with the major agencies and states affected. In the early days, public acceptance was facilitated by the fact that the federal government provided funding for most of the projects constructed.

Computers and Modeling

As in many other aspects of modern life, computers have become an indispensable tool in water resources planning and management. Capabilities of computers are increasing year by year. After preliminary work has been done, computations that once would have needed months for a staff to accomplish can now be produced in days or hours. A good example is the simulation of water quality processes in the Chicago waterway and upper Illinois river systems (Macaitis, *et al.*, 1993). The traditional optimization of multiple objectives can be expedited by use of computers. A good review of the use of computers in water resource planning has been published (Grigg, 1985).

In our enthusiasm for computers, we should not overlook the fact that they can only give complete answers when all the variables in the equation can be quantified. This limitation is of great importance in the new era, when so many different agencies become involved, and environmental and other intangible values may be determinative. The use of computers and modeling can be adapted to these more complex situations by special means explained in Chapter 11.

Present Situation

In the United States today, almost all sizable rivers are already developed for one purpose or another. Problems of operations and maintenance and maintaining flow for existing structures are relatively more important than new projects. New planning initiatives usually take the form of adjusting and modifying existing systems, rather than developing a completely new plan. Traditional multiple-purpose planning is becoming increasingly inadequate to deal with water problems now arising. It concentrates too much on maximizing national economic benefit from the construction projects. Environmental values, other intangibles, and states' interests are brought into the planning process too late for those interests to feel that they are a part of the process. The planning of resources development should recognize broader objectives, including pollution control, runoff control, and the environment. The continued growth of our national economy requires a corresponding increase in consumptive use of water for water supply and accompanying industry. If unrestrained, this results in a continuous accentuation in competition for water and general dissatisfaction by environmental interests. The following chapters discuss how such problems may be handled.

References

Grigg, N.S. "Water Resources Planning." McGraw Hill, 1985.

Macaitis, B., Variakojis, J., and Mercer, G. "Simulation of Water Quality Processes in the Chicago Waterway and Upper Illinois River Systems," in *Water Management in the '90s*, ASCE, 1993.

Chapter 8

WATER QUALITY AND POLLUTION CONTROL

Introduction

In the past, water pollution control was not a part of multiple-purpose planning for two reasons. In the first place, there was no active federal pollution control program until after the Clean Water Act of 1972. In the second place, after the federal program was initiated, it first mainly consisted of standards for treating effluents from municipal waste treatment plants and major industries, without apparent relationship to the programs and objectives of the construction agencies.

EPA programs are now being extended to include other major pollution sources, as indicated in the following chapter. Also, major water resources planning objectives have been extended to include fields in which water quality is important. Water quality, especially toxics, is a key influence on environmental problems in many rivers and lakes and all estuaries, and in groundwater along the coasts. Nutrients also cause a good many problems. Nutrients in groundwater are dangerous to human health, especially to pregnant women and newborn babies, as indicated in Chapter 4. In lakes and estuaries, nutrients encourage photosynthesis and excessive algal growth, which are very detrimental. Nutrients are persistent in that they are not appreciably reduced by ordinary detention storage. The National Estuary Program has to consider such aspects as a matter of key importance.

Nonpoint Sources

It is now generally acknowledged (as researchers have known since approximately 1970) that a large part of the pollution in our streams comes not from well-defined municipal and industrial wastes, but from nonpoint sources such as homes, commerce, roads and streets, agriculture, and gasoline filling stations.

Major industrial facilities have long had their wastes controlled under the Clean Water Act. Figure 8-1 shows such a facility. It does not take much of an imagination to realize that all of the accompanying activity, truck traffic, and materials handling will inevitably result in heavy runoff pollution whenever rain occurs.

Figure 8-1. Industrial Source of Runoff Pollution, Raritan River. Source: Rutgers University and Will Gainfort.

In view of the diversity of the nonpoint sources, as discussed in the following chapter, control can seldom be directly related to water quality objectives.

A new need for nonpoint source pollution control has recently arisen from a new direction. For maintenance of navigation along our coasts and rivers, and in harbors, an increased amount of dredging is required, of which 5-10% is contaminated, including nonpoint-source material. This amounts to 14-20 million tons of contaminated material annually. In order to avoid treatment costs, sometimes this material can be used for construction of islands, or foundation for habitats. This problem has been studied by the National Research Council, which considered that states and the EPA should take action to control the upstream sources of pollution (Anon., 1997), but a major long-range problem remains.

Water Quality Criteria and Standards

The water quality in our rivers, streams, lakes, and estuaries is presumably the main objective of EPA policy. Control is needed not only for protection of the plants, fish, and animals of the environment, but also for the public health of humans. Under the Safe Drinking Water Act, standards (called "maximum contaminant levels") are required to be met by drinking water. The Clean Water Act reduces discharges of pollutants into streams, and the Safe Drinking Water Act establishes levels that are the maximum allowable in potable water, with the states, municipalities, and water companies being responsible for the end result. Also, water quality criteria for ambient waters are established and are intended to protect aquatic organisms. Water quality standards and criteria have been widely criticized as not being scientifically well based (Abelson, 1992; Ames, 1994; Savage, 1994; Wheeler, 1994), but they are obviously necessary, and presumably will be continued and improved.

Although the EPA programs for treatment of effluents have been a positive and generally successful factor, the new EPA initiatives for municipal and industrial control are far from clear as to their specific intentions and their realism, and they do not cover all dispersed sources. In order to achieve all desired objectives, they must be considered together with programs of the states and other agencies. The newly emerging program of establishing total maximum daily loads is even more uncertain, as discussed in Chapter 9.

All told, there are not less than seven EPA programs that may deal with nonpoint-source pollution:

- Municipal controls
- Industrial controls
- Coastal zone controls

- CSO controls

- National estuaries program

- Total maximum daily loads

- Groundwater controls

In each program a separate technical approach is specified. An industrial facility may be subject to several entirely different regulations covering, a municipal permit, an industrial permit, a coastal zone permit, and compliance with total maximum daily loads. Obviously, this multiplicity of approaches is undesirable. The EPA, hoping to integrate results of its various programs into a common runoff control (wet weather) approach, is considering a framework for "watershed permitting." This framework would highlight the most critical pollution control needs of a watershed or community and allow them to be addressed in the most effective and least costly manner. Areawide pollution control planning was authorized by Section 208 of the Clean Water Act and was actively carried out for some years, as indicated in Chapter 1. Under present conditions, it would certainly be desirable.

Conclusion

One of the key problem areas in water resource planning is pollution. The basic programs under the Clean Water Act that control wastes from major industries and municipal waste treatment plants are generally successful and are adequately handled by EPA and the states. Standards are based on technology. There is no reason why this program cannot be completed throughout the United States, without special interagency coordination or basin planning. Of course, the results are not yet completely satisfactory. Some pollutants remain that may require special treatment for use in water supply. Also, the water quality standards may not be scientifically precise and sufficient to meet all ecological needs downstream, but the EPA is working on refinements.

The pollution control program as a whole now involves new aspects that cannot be as independently conducted as the original programs were. Most of the problems requiring close coordination with other agencies and states involve pollution from nonpoint sources and runoff, and those sources affecting groundwater. They are discussed in Chapter 9. These aspects must be taken into account if we are to solve the difficult problems that are now arising.

The new EPA initiatives to reinstitute areawide planning of pollution control would still leave unsolved basic problems of coordination with other agencies and between states.

References

Abelson, P.H. "Exaggerated Carcinogenicity of Chemicals, *Science 236*, 609, 19 June 1992.

Ames, B.H. "Does Current Cancer 'Risk Assessment' Harm Health?" in "National Water Resources Regulation: Where Is the Pendulum Now?", American Society of Civil Engineers, 1994, p. 121.

Anon. "Contaminated Sediment Management Needs Better Assessment Tools, N.R.C. Says," *Environmental Science and Technology*, 31, 5, 1997.

Savage, R.H. "Clean Water Act Reauthorization: The States' Perspective," in *Water Resources Update*, Universities Council on Water Resources, No. 94, Winter 1994, p. 28

Wheeler, D.R. "Where Is the Pendulum Now?" in "National Water Resources Regulation: Where Is the Pendulum Now?", American Society of Civil Engineers, 1994, p. 155.

Chapter 9

RUNOFF POLLUTION CONTROL/STORMWATER MANAGEMENT

Introduction

Runoff pollution control deals with the problems of controlling nonpoint-source pollution (NPS)* along with erosion and headwaters flooding. Originally the term *stormwater management* applied to localized control of floods. It is now becoming good practice to combine the control of floods with the reduction of pollution of floodwaters, designating the combined structure as a dual-purpose stormwater detention basin and the program as runoff pollution control.

Control of Runoff Pollution

Control of runoff pollution is a subject that until recently has been somewhat neglected by federal planners because of the widely dispersed sources. As previously discussed, basin-wide plans to curb pollution effectively, either ground or surface, either for water supply or for general environmental purposes, must include control of runoff pollution. This is a program not very appealing to bureaucrats because there are no large conspicuous targets, such as major industries and sewage treatment plants, but rather widely distributed elements of the voting public, including farmers, small businessmen, and home owners, with their garbage cans, pets, and automobiles. These numerous offenders cannot be controlled directly by water quality standards or criteria. The number of points of control is generally too great for treatment to be an option.

---

* *This book refers to nonpoint pollution in the usual sense of dispersed sources. Unfortunately, the Clean Water Act classifies as nonpoint sources only that pollution that does not reach a stream through a pipe, ditch or channel.*

Particularly in the first part of a storm, the concentrations of pollutants in runoff range widely, rising rapidly to a short peak and generally diminishing more slowly but irregularly thereafter. In such situations, it is impossible to state what particular level of pollutant controls the damage to the various species of aquatic organisms, particularly as part of the damage done may result from excess scour, sedimentation, and habitat disruption rather than the toxicity of pollutants.

Almost any kind of pollutants can occur in stormwater. Where nitrates or other nutrients are found, there is usually no remedy but control at the source. When ingested, nitrates are extremely harmful to babies and pregnant women, and in lakes or estuaries nitrates can result in excess algae growth. Algal blooms can cause extensive fish kills through toxins and by causing anoxia; they may also cause shellfish poisoning (Vitousek, 1997). In some cases, as in the St. Johns River in Florida, excessive algal blooms caused by nutrients reduce the light necessary for survival of eel grass. Young fish of many species that shelter in eel grass are adversely affected by the loss of their habitat (De Sena, 1997b).

Many nonpoint-source pollutants are controlled best by management practices, such as detention basins, while others can be controlled at source. With this complex set of relationships, generally orchestrated but not entirely controlled by the EPA, the construction agencies can

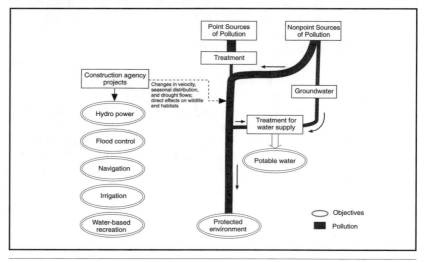

Figure 9-1. Pollution Control and the Environment.

have no direct control over water quality. However, the effects of the construction projects interact with the pollution from both point and nonpoint sources to affect the ecology of streams and estuaries. This relationship is roughly indicated by Figure 9-1.

This figure illustrates that both human and non-human environmental aspects require equal control of nonpoint sources and point sources, and that the remaining pollution and the construction agency programs have a combined effect upon the environment.

When to Control

As a matter of priority, nonpoint sources should be controlled where they impact environmentally sensitive areas, coastal waters, watersheds, aquifer-recharge areas, water-quality-limited stream segments, or well-head-protection areas. However, since water pollution is a national problem, it is always desirable to control pollution from nonpoint sources.

One practical aspect of runoff control is particularly important to policy. It is almost invariably much more difficult and expensive to reduce pollution in the runoff of existing facilities than it would have been when they were first built. Detention basins can be readily included in the layout of a new development at moderate cost. After the facility has been built, however, it is often impossible to provide stormwater management at site, and treatment facilities downstream are sure to be more expensive. Hence the policy: as a first priority, incorporate runoff control facilities in all new developments, but provide control for runoff from existing facilities only after careful planning and analysis.

Nonpoint-Source Pollutant Removal Methods

Since nonpoint-source pollution originates in such numerous and widely dispersed locations, and is not readily adaptable to classification by numerical criteria, it must be controlled mainly by best management practices (BMPs) applied either at source or through control structures. As regards agricultural nonpoint-source pollution, control at source is usually considered the most effective. However, according to a GAO study, where agriculture is the main source, a watershed-based approach, both federally and locally backed, and considering both surface and ground-water quality, is considered essential to success (Anon., 1995). As regards methods, manure and nutrient management are essential, along with exclusion of livestock from streams, and the enhancement or replacement of riparian wetlands (Brenner and Mondo, 1995).

Common pollutants in NPS runoff include the following: total sediment, BODs, lead, nitrogen compounds, and cadmium. In addition, hydrocarbons in considerable quantities are present in runoff, especially from streets, highways, and filling stations. They are particularly harmful in chlorinated drinking water. They are also harmful to some organisms, specifically oysters (Haskin, 1987).

A leading water company executive writes: "Non-point pollutants (e.g., oils, road salts, microbiological contaminants, VOCs, SOCs, humic and fulvic acids) are the things that seem to be driving water treatment plant designs today. It may be appropriate to draw a distinction between what we have already accomplished in regard to point-source pollution control and what remains to be done regarding non-point sources" (Howard Woods, New Jersey American Water Co., personal communication).

The metallic pollutants, hydrocarbons, sediment, and BOD are largely removed from runoff by impoundment of 24 hours. Both the BOD components and the hydrocarbons are biodegradable and therefore do not accumulate in a detention basin unless present in abnormal quantities, and the metallic pollutants are usually associated with sufficient sediment as not to cause a maintenance problem when removed.

Although the usual method of removing runoff pollution is by detention basins, other methods are sometimes employed. In one case in New Jersey, control outlets were provided for two existing swamps, which served very effectively to remove pollution. More recently, stormwater-treatment wetlands have been constructed in a few cases (White and Meyers, 1997). These are essentially very shallow wet basins that allow the free growth of typical wetland vegetation. In addition to the settling processes of any dual-purpose detention basin, the wetlands vegetation and the resulting detritus absorb considerable pollution through adsorption, bacterial action, and nutrients for plant growth. Surprisingly enough, in a monitored case, removal of phosphates by wetlands was greater than that of nitrates. Stormwater-treatment wetlands, although environmentally very beneficial, have two serious handicaps: they take a great deal of land, and they are an ideal breeding ground for mosquitoes.

It so happens that it is relatively inexpensive to apply dual-purpose stormwater detention to new developments, whereas it is much more expensive, and sometimes entirely impracticable, to provide equivalent NPS controls for existing facilities. This leads to an important conclusion. In basin-wide water resource planning, dual-purpose stormwater management should be required for new or rebuilt facilities generally throughout the United States, except individual single-family dwellings, while most other measures of flood control and pollution control should be carefully planned and developed to meet the needs of each river basin.

Design of Dual-Purpose Detention Basins

The BMP for dual-purpose detention basins is a simple and widely applicable technology-based approach. As shown in Figure 9-2, a frequently occurring storm is designated, called the water quality design storm. The (lower) portion of storage, which holds the runoff from the water

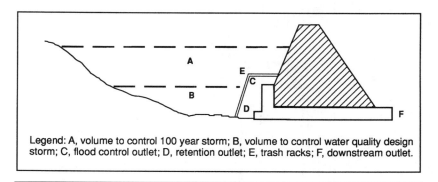

Legend: A, volume to control 100 year storm; B, volume to control water quality design storm; C, flood control outlet; D, retention outlet; E, trash racks; F, downstream outlet.

Figure 9-2. Dual-Purpose Detention Basin.

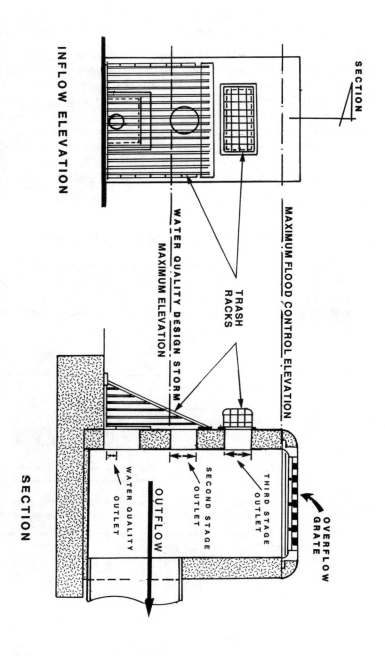

Figure 9-3. Dual-Purpose Detention Basin Outlets.

quality design storm, is drained by a small fixed outlet that is designed to evacuate the runoff accumulation from this design storm, usually in about 24 hours. The upper portion of the impoundment is designed to store runoff from a larger storm, for control of damaging floods. It is provided with a larger second-stage and possibly a third-stage outlet. (See Figures 9-3.)

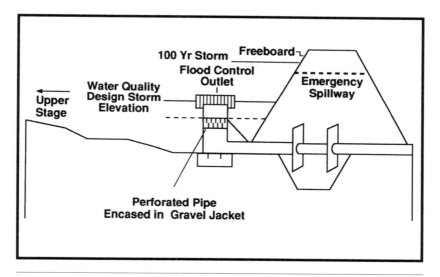

Figure 9-4. Simplified Dual Purpose Detention Basin.

A spillway must be provided to ensure safety of the structure in even greater floods. For the smaller detention basins simple pipe outlets may be provided, with perforated pipe in place of a water quality outlet, as shown in Figure 9-4. In either case, a trash rack of some kind is ordinarily installed before each outlet.

In many cases, the same detention arrangements are built with extra space below the level of the outlets, so as to provide a wet basin, which is environmentally beneficial as well as being more attractive in a housing development. In some cases, infiltration basins are also used, allowing groundwater to be recharged as an additional benefit. Upstream of the detention basin, steps may be taken to limit erosion and retain sediment by grass strips, swales, and other soil conservation techniques.

The technical aspects of stormwater management impoundments are now well understood. They have been included in a manual by the Metropolitan Washington Council of Governments (Schueler, 1987) and more recently in a Manual of Practice, No. 77, ASCE (ASCE, 1992). However, where state or federal regulations and criteria include design criteria, they must be followed.

Regional Detention Basin Systems

Regional planning of stormwater management is almost always more advantageous than the system of requiring each development to build its own basin. One characteristic of detention basins that greatly favors regional planning and regional detention basins is the rapidly decreasing effect of short-term detention on flooding downstream (American Society of Civil Engineers, 1992). Detention basins designed to reduce flood peaks at site are largely ineffective a few miles downstream. Therefore, the "regional" basins are more effective for flood control downstream than numerous small basins at site. As concerns the removal of particulates, the larger regional basins are more economical to construct than numerous small basins. For both control of flooding and runoff pollution control, the planning must carefully balance the needs of the immediate vicinity with the objectives further downstream.

Regional systems require more careful planning, and provisions for funding are not automatic, as the need for funds for the larger regional basins does not usually correspond to the availability of funds through fees charged to builders. Also, the larger regional basins are more apt to encounter difficulty of encroaching on wetlands. With these problems, regional systems are usually developed in one of three ways: by large landowners on their own land, by municipalities or counties, or by stormwater utilities authorized by state legislation.

Federal Requirements

One federal regulation, the future interpretation of which will be troublesome, is the expressed objective to reduce pollution from municipal runoff "as far as practicable." An official interpretation was made at one time that this means compliance with numerical water quality standards in the cities designated under this program. To require this for an existing sizable municipality would be completely unreasonable. The runoff from municipalities comes largely from streets and minor nonpoint sources, and it is difficult to establish in a rapidly varying outflow what level of pollution should be taken as representative. It seems that watershed analysis would be essential to determine what sort of controls are appropriate for municipalities in each particular case. This would require a broad approach. Presumably, as time goes on, the EPA will arrive at some more practicable criterion.

The new program of individual permits for separate industries is of great potential scope. In New Jersey alone, there are 28,000 facilities that might require permits. The ultimate standard that runoff and discharges from each site will be required to meet has not been established. Interpretation of the loose language of the law could make the standards very burdensome. However, initial requirements are not onerous.

Another federal requirement that will cause difficulty is the newly developing program of "total maximum daily loads" (TMDLs) (Chapter 1). The inherent problems are very great. Since the greater part of present-day pollution comes from nonpoint sources, the institution of a TMDL for any given substance at a given point of a river imposes a responsibility on the state (or the EPA) to control pollution from NPS (and all other sources) in the drainage basin above. Probably no state has statutory authority to control all nonpoint sources, including automobiles, individual homes, individually owned lawns and gardens, and agriculture. Also, the use of TMDLs for rivers draining more than one state present additional difficulties.

The law implies that achievement of water quality standards in the streams is the objective of TMDLs. We may assume that EPA officials will act with discretion in imposing such controls, but it will be technically difficult to avoid imposing standards that will be difficult to administer. To add to the problem is the fact that low flows of water in times of drought, or when depleted by water rights, may result in environmental harm from pollution permissible under a TMDL that is quite acceptable during normal flow conditions.

State Programs

Many of the states have adopted useful stormwater management programs with controls for pollution and storm runoff. Maryland requires approval of a storm water management plan for each new development, with a general goal that both the quantity and quality of runoff from developed land shall be as close as possible to the characteristics of the area prior to development. There are tighter controls on areas closer to Chesapeake Bay, designed to make a major reduction in inputs of nitrogen and phosphorous to that estuary, in accordance with objectives of the National Estuary Program. State grants are available to help with the funding (Whipple, *et al.*, 1992).

Florida was one of the first states to control nonpoint-source pollution. All development and redevelopment projects are required to obtain stormwater permits. This program is being delegated to the state's five regional water management districts, although the process of delegation is not yet complete. The state policy is that post-development stormwater peak discharge rates, volume and pollutant load should not exceed pre-development conditions. There are also spe,cial goals for runoff from inadequately treated agricultural and urban discharges and runoff entering fishable and swimmable waters. Local governments are required to prepare watershed management plans. Matching grants are available for stormwater quality projects (Whipple, *et al.*, 1992).

Wisconsin is another state that has made major progress in runoff control. By 1989, remedial programs of NPS control had been undertaken on 31 priority watersheds. The original program was directed largely at reducing existing agricultural pollution of lakes and trout streams. Governmental cost-sharing has been available in most cases, but landowner participation is voluntary. Wisconsin has not yet attempted to achieve a similar success in its urban and urbanizing areas (Whipple, *et al.*, 1992).

California has a well-developed stormwater control strategy, based upon basin plans for the various watersheds. It is administered through nine regional control boards. However, as in other states with dry climates, grass strips and swales are generally inapplicable, and dry detention basins are not desirable (Whipple, *et al.*, 1992).

Delaware has a statewide program requiring water quality control of storm runoff from new development. A key aspect is that efficiency of removal of suspended solids must be 80%. The water quality design storm for pollution control is the two-year storm. Use of wet ponds is encouraged. Delaware also is subject to controls under the Delaware Estuary Management Plan (Delaware, 1991).

New Jersey has well-established standards for runoff pollution control, which, however, until recently, have been applied mainly at the discretion of municipalities.* However, in 1997, a new directive, "Residential Site Improvement Standards," provided standards for residential developments applicable statewide (New Jersey, 1997). This requires that the peak flow downstream from 2-, 10-, and 100-year storm events after improvement shall be no greater than flows that occurred prior to the construction. For water quality, the water quality design storm is a 24-hour, one-year frequency, and it shall be 90% evacuated (from dry detention basins) in not less than 18 hours. New Jersey is also subject to controls established under the Delaware Estuary Management Plan.

Some states, such as Pennsylvania, have no statewide program of runoff pollution control, although they may have programs to protect lakes or other particularly sensitive areas.

The technology-based programs of the numerous states to control nonpoint-source pollution, while differing in detail, are generally realistic and effective for new construction. They are in stark contrast to the sweeping mandates under the EPA TMDL program and the new municipal and industrial controls, which require a result without regard for the practicality of implementation. This matter is further considered in Chapters 10 and 11.

Runoff pollution control on new development costs money, but it is not without economic justification. Developers and businessmen interested in the growth of the economy naturally would like to avoid the expense of building detention basins and other features to minimize runoff volume and pollution. That is the reason why so many states still do not mandate adequate measures with every development. However, it must be borne in mind that even the best runoff control measures do not remove *all* of the pollution. New developments, in paying for runoff control measures, are minimizing but not entirely eliminating the accentuated flooding and added pollution caused by the development. From a practical viewpoint, this is all that can be expected of developers, and it is always justified. Those who oppose mandatory runoff pollution control are in effect advocating that the state assume the burden either of remedying such effects after construction or of suffering the increase in environmental degradation resulting from the additional flood runoff and pollution. They are laying a burden on the future.

* *Also in areas draining into or under the Delaware and Raritan Canal.*

References

American Society of Civil Engineers. "Design and Construction of
   Urban Storm Water Management Systems." ASCE Manuals and
   Reports of Engineering Practice No. 77, Chapters 11 and 12, 1992.

Anon. "Watershed-based Approaches May Be the Solution to US NPS
   Pollution," *Environmental Science and Technology*, vol. 29, 9, Sept.
   1995.

Brenner, F.J. and Mondo, J.J. "Nonpoint Source Pollution Potential in
   an Agricultural Watershed in Northwestern Pennsylvania," *Water
   Resources Bulletin*, 31, 6, 1101, Dec. 1995.

Delaware, State of. "Delaware Sediment and Storm Water Regulations." Jan., 1991.

De Sena, M. "Decline of St. Johns River Traced to Nutrient Overload,"
   U.S. Water News, 14, 8, p. 1, Aug. 1997b.

Haskin, H. (1987). "Petroleum Industry in the Delaware Estuary," a
   report to the National Science Foundation by Whipple, W. and
   Patrick, R., RANN Program. Grant No. ENV. 74-14810-A03.

New Jersey, 1997. Administrative Code, Title 5, Chapter 21, 1997.

Schueler, T.R. "Controlling Urban Runoff," Metropolitan Washington
   Council of Governments, 1987.

Vitousek, P.M., Mooney, H.A., Lubchenco, J., Melitto, J.M. "Human
   Domination of Earth's Ecosystem," *Science*, 277, 494, 25 July 1997.

Whipple, W., Berg, V.H., and Livingston, E.H. "Statewide NPS Strategies," *Water Resources Planning and Management*. ASCE, 1992.

Chapter 10

COORDINATION UNDER PRESENT INSTITUTIONS

Introduction

In recent years, developing our traditional water resources has been increasingly difficult because of declining opportunities, environmental roadblocks, and increasing contention over water. Our country has continuing needs for navigation and hydroelectric power, variable needs for irrigation, and increasing needs for water supply and flood control. However, the environment has suffered various sorts of damage and loss, some of which results from water resource development. There are a great variety of government organizations charged with various aspects of these water and environmental fields. Because of the increasing scopes of the problems, more and more of the functions involved are now the responsibilities of states. However, there is no way that responsibilities for water planning could simply be delegated to the states. It remains true that the federal agencies usually have larger staffs and greater expertise in technical matters than their state equivalents. What is even more important is that most large watersheds include more than one state, so that a federal or interstate agency can more effectively solve the problems.

Besides federal and state agencies, with their varying objectives, there are many other stakeholders, including industries and environmental organizations. Inevitably, there are different viewpoints. Coordination attempts to find some way of reconciling these diverse interests or of merging them into a single conclusion. Of course, in the past, this has long been done in a number of ways, but increasingly, the divergences are becoming more numerous and intractable. Accordingly, the problem of coordination is becoming more serious.

In proceeding with coordination, the question must be answered as to which functions can be appropriately left to individual agencies or states, and which will require improved federal interagency action. Some functions can be technology based and applied generally in all regions, but others must be planned and designed to meet the needs of a particular region or watershed.

There are many parts of the country where the water resources situation badly needs serious restudy to resolve major differences. Public officials, whether state or federal, should be aware of this and should take any opportunity to work out some kind of cooperative effort within the present institutional framework. This is more difficult than it would be with full congressionally sanctioned arrangements for cooperation, but in some cases it is still possible.

Basic Policies for Coordination

Major programs of basin-wide flood control, hydroelectric power, navigation, water supply, and irrigation need coordinated planning and management. Although agencies concerned can be relied upon to design and build the projects, there are usually potential problems with environmental aspects (both favorable and unfavorable), and frequently differences of interest between stakeholders or responsible agencies. Several important policies should guide and assist this planning. The first is environmental preservation. Although we cannot accept it as an absolute mandate, the protection of endangered species (and other watershed ecology) is an important national objective. Second, structural provision for flood control should always be accompanied by floodplain management, and part of floodplain management should be preservation of a strip in the floodplain for fish and animals and wetlands. Third, the benefit/cost ratio is (still) of major importance. This is the ratio between the national economic resources created by the project and the national economic resources utilized. Protection of human life, liberty of action, and welfare of all concerned is also important, as explained in Chapter 1. There are, of course, in addition, the usual objectives of avoiding harm to religion, archeological relics, the customs and beliefs of Native Americans, and so on.

Controversial Aspects

Adherence to these main policies will reduce the potential areas of controversy but will by no means eliminate all of them. There will remain the entirely legitimate contentions between areas, each needing more water for a growing population. There will remain problems where large dams, particularly those generating hydroelectric power, harm fish or wildlife (e.g., Columbia River and Colorado River). There will remain cases where upstream states need more water for growing populations and irrigation while downstream states need the water for endangered species, wetlands, fishing, and/or oysters. The Alabama-Coosa-Tallapoosa and Apalachicola-Chattahoochee-Flint River Basin Combined Study is such a case. There are many other major rivers where major needs for coordination exist, due primarily to the differing needs of the different populations involved, and the objectives of the participating state and federal agencies.

Government Attempts at Coordination

At present there is no federal agency officially charged with such coordination, although some laudable attempts are being made. River basin commissions can play a useful role in planning and coordinating agency and state action within the basin. However, river basin commissions cannot provide a complete answer. Whether or not constituted with a federal chairman, the commission has insufficient authority over states and the federal agencies. Besides, only a few of such commissions still exist.

The EPA has successfully headed some important and generally successful studies under the National Estuaries Program. These studies are important, but their major objective is simple: preserve the estuary and bay from pollution, especially nutrients. The main participants are the EPA and the adjoining states. It is important in such studies that the states have a direct interest in both the objective of the study and also of the principal means of achieving it, namely, the control of runoff pollution from development and from agricultural lands draining to the estuary. Other water interests, such as flood control, hydroelectric power, water supply, and navigation are not much involved. Thus, this program does not indicate a pattern for the other complex problems of interstate coordination.

As seen in Chapter 2, there are some real opportunities for cooperation between construction agencies, states, and environmental interests in restoration of wetlands, especially where large amounts of fill are available from dredging, and also where there are environmental interests of great concern, especially the Florida Everglades. There are other cases where flood control and erosion control work can be planned so as to provide wildlife habitat along the banks of streams and in wetlands.

Problems involving harmful effects of construction on wetlands can be greatly eased by wetlands "banking," which allows important construction to be undertaken contingent upon provision of replacement wetlands elsewhere. All of these programs are useful, but they cannot be extended to cover the majority of difficult relationships between environmental regulations and the planning of water resources, especially where differing agency viewpoints and two or more states are involved.

Cal-Fed Bay Delta Program

A notable attempt at federal-state coordination is the Cal-Fed Bay Delta Program. This is an ambitious program designed to provide better management in the area of the Sacramento/San Joaquin Bay Delta area. It was initially funded mainly by the state of California. Principal concerns are protection of fish and wildlife, water supply reliability, vulnerability of levees, relief from natural disasters, and water quality. Joint direction of the effort is by the governor of California and the Secretary of the Interior, but numerous state and federal agencies participate. The Cal-Fed is potentially a very powerful arrangement, but only one state is involved. Therefore, some more complex arrangements must be made where the region includes more than one state.

The Alabama-Coosa-Tallapoosa and Apalachicola-Chattahoochee-Flint (ACT-ACF)

These river systems have long been the subject of controversy. In 1983, in order to avoid litigation, certain preliminary steps were agreed upon. However, the controversy widened, and in 1990, Alabama filed suit against the Corps of Engineers, challenging the proposal to reallocate a part of flood control storage to water supply. This lawsuit challenged the proposed reallocation on two counts: first, because downstream would ultimately need the water, and, second, because environmental interests downstream would be adversely affected. Florida and Georgia formally intervened in the suit. In 1992, it was agreed by all concerned to hold the lawsuit in abeyance until a new comprehensive planning effort could be carried out. An appropriation of 12 million dollars was made for this study. The completion of the study was scheduled for 1995 and later extended to 1996. In the conduct of this study, the Greeley Polhemus Group was retained by the Corps of Engineers as consultants, with the effort being headed by the author. In 1996, a draft completion report of that study was issued (Werick, 1996). In spite of prolonged efforts, no definite agreement resulted from the study. However, it is noteworthy that the operational alternatives of operating the nine federal reservoirs in the two rivers were studied, along with an analysis of water supply, including water-demand forecast until the year 2050. The studies covered water-related recreation, river water quality, irrigation, navigation reliability, interbasin transfers, and the environmentally significant effects in Apalachicola River and Bay. This complex analysis was conducted with the latest methods of cooperative computer studies, more fully described in Chapter 11.

There have been recent attempts to revive the processes of coordination for these basins. Under a tri-state agreement reached in 1996 (among Alabama, Florida, and Georgia), two interstate compacts would be formed under which participants would have until the end of 1998 to reach agreement on allocation of water from the two river systems. Federal legislation recently introduced would create these two separate compacts, one among all three states to allocate water from the Chattahoochee, Flint, and Apalachicola rivers, and one between Georgia and Alabama to allocate water from the Tallapoosa, Coosa, and Alabama rivers. Each compact would be administered by one representative from each state, plus a federal representative appointed by the President. Each state would have veto power over any compact decision (Anon., 1997). In principle, these compacts should have a reasonable chance of success because the completion of the cooperative computer studies and evaluation of alternatives remove most of the usual technical obstacles towards arriving at a solution, and the single federal representative could coordinate differences of viewpoint between the federal agencies represented.

Platte River

An important example of coordination with existing institutions is the recent agreement between the Interior Department and three states regarding the Platte River. This agreement involves mainly reconciliation of needs for water supply and irrigation with preservation of wildlife, especially the whooping crane and other endangered species. The agreement calls for a comprehensive basin-wide environmental study of the Platte River. Under the agreement, each state will spend $200,000 annually for three years, while the federal government will spend $2,500,000 (De Sena, 1997a). This is a promising beginning, but it remains to be seen how successful it will be.

Water Pollution Control

One of the key problem areas is water pollution. The basic programs under the Clean Water Act, which control wastes from major industries and municipal waste treatment plants, are generally successful and are adequately handled by EPA and the states. Standards are technology based. There is no reason why this program cannot be completed throughout the United States, without special interagency coordination or basin planning. Of course, the results are not yet completely satisfactory. Some pollutants remain that may require special treatment for use in water supply, and the water quality standards may not be scientifically precise and sufficient to meet all ecological needs downstream, but the EPA is working on refinements. However, the control of nonpoint-source pollution remains a major problem nationwide.

Conclusion

When situations become intolerable, some means will be sought to obtain coordination. Where only one state is involved (Cal-Fed), or where the interstate planning and modeling has been accomplished by special means (ACF), it will be easier to find solutions. In some circumstances, cooperative efforts have been successful. However, some of the arrangements are complex, and it remains to be seen how well they will work. If no other solution is found, recourse may be had to long and expensive litigation and decisions of the Supreme Court. This is a prolonged process that can cover only a few of the outstanding issues. It should be possible to find a better holistic approach to using our waters to the benefit of all concerned, while still preserving the environment.

References

Anon. "Southeastern Senators Introduce New Interstate Water Compact Bill," *U.S. Water News*, 14, 8, 8 August '97.

De Sena, M. "Governors Sign Three-State Platte River Agreement," *U.S. Water News*, 14, 8, 1 August 1997.

Werick, W.J., Whipple, W., Jr., and Lund, J. "Basinwide Management of Water in the ACT and ACF River Basins," Draft Report to the States and the Corps of Engineers, August 1996.

Chapter 11

COORDINATION THROUGH FED-STATE PLANNING

Introduction

If more effective coordination needs to be developed among key federal agencies and the states, the question is how?

Some of the difficulties could be faced by creating another TVA for each major river basin nationwide, but this old idea has long since lost its appeal. However, it would be equally unreasonable to expect that some federal superagency could be created to assume responsibility for all of these diverse functions. More powerful federal government is not the answer. Our country has moved beyond the stage at which federal agencies could plan and build entire river systems with only nominal cooperation from state and local agencies. In order to devise a more effective arrangement, we must first consider how to surmount the technical difficulties of obtaining understanding among different interests in complex studies.

Computers in Fed-State Planning

In complex arrangements involving a number of federal and state agencies, usually with some conflicting interests, the use of computers can be a great facilitator. An effective approach to using computers in multiple-agency planning situations has been developed and applied by researchers at the University of Washington (Palmer, *et al.*, 1993). This technique, known as "shared vision modeling," was developed during the National Study of Water Management during Drought and applied extensively in the ACT-ACF Basinwide Study. A shared vision model is a highly interactive model of resource conflicts, in which managers, operators, and stakeholders are actively involved in the development as well as the use of the model for decision making. (Stakeholders are entities outside of government, such as power companies or water companies, that may have expertise and an interest in the study.) The model uses graphically based computer simulation to develop easily understood analyses of the systems under study and facilitates the testing and collaborative use of the model by all those involved in the process. When applying such a modeling approach, care is taken to ensure that the study objectives are reflected in the model so that the issues that will affect decisions are incorporated at an appropriate level of detail. The questions of who will use the model and how it will be used help define the appropriate level of model complexity. The advantage of these "shared vision models", as the name implies, is that consensus in the model and in the computer results can be reached, since all parties participated in the development of the model.

In sizable river basins it is common to have a variety of uses of water, each with its own particular interests and stakeholders. A variety of alternatives may have to be considered, such as:

- More (or less) water for municipal water supply
- More (or less) water for irrigation
- Strict (or lenient) water conservation measures
- More (or less) fluctuation in power plant releases
- More (or less) flow in navigation channels
- More (or less) fishing and lake recreation
- More (or less) storage reserved for flood control (or power)
- Possible additional storage for supply of water during drought
- Availability of water during low flow periods in estuaries and wetlands, and for endangered species
- Water pollution control alternatives

All of the above alternatives may need to be considered with varying conditions of flood and drought, with projections including future changes due to population and industrial developments, and with consideration of groundwater, if applicable.

With such a wide variety of technical questions coming to bear on defining the system under study and how it responds to alternatives, it is apparent that any single vision of the system would be incomplete and that no single agency working in isolation would be able to obtain agreement on the output of the model after the fact. However, with a well-prepared shared vision model, all of the technicalities can be ironed out and agreed upon in advance, and analysis of the various alternatives becomes relatively simple.

There are, of course, some qualifications in this optimistic picture. First, in principle, modeling can only be accurate if all the variables are tangible and quantifiable. The models may be able to predict, say, the increase in salinity in Apalachicola Bay caused by allocating 10% more water to irrigation during a 10-year-frequency drought, but they have no way of evaluating whether the deleterious effects upon the various species of wildlife and plants involved is more important than the extra profits to irrigators in the headwaters.

Secondly, all of the modeling costs money, and someone must take the lead in organizing it and getting the money. Some lead agency is required for the modeling. In the ACT-ACF study the modeling work was well organized by the Corps of Engineers and was successful, despite the widespread differences in objectives between the federal and state agencies and their respective stakeholders. The failure (so far) of the ACT-ACF effort to obtain a final plan and the limitations of the results as regards evaluation of intangibles do not affect the validity of the shared vision modeling concept. Although it is not a cure-all, shared vision modeling is a prerequisite to successful handling of the vastly complicated problems that characterize the new era. The tough decisions still must be made as to giving advantages to human development, while minimizing damage to the non-human environment. One way to do this is covered below.

The Fed-State Study

The multi-agency preparation of complex state-federal basin-wide studies under federal leadership would require congressional authority and funding. It would also be necessary to give clarification as to the nature of some of the absolute powers of EPA. The Corps of Engineers, or, in appropriate parts of the country, the Bureau of Reclamation or the TVA, could perfectly well take the lead in bringing the states and other agencies together for such studies.

These studies would be conducted under congressional authority, whereby the convening agency would arrange cooperative arrangements with other federal agencies and the states. After preliminary meetings, arrangements would be made for computer experts from the agencies and states to work together. The various alternatives would be explored and evaluated, and the key remaining questions outlined. The states and all the federal agencies would be allowed (and required) to publicly express their recommendations on the controversial questions. If large economic benefits were to result from the taking of certain wetlands or endangered species habitat, the states concerned, the EPA, and the Fish and Wildlife Service would have to recommend, publicly, which course was preferable. Of course, some way must be found to settle differences of opinion. Adjustments can often be made whereby a technical solution can minimize potential political problems, but ultimately decisions must hold the balance between human development and ecological conservation and decide whether the interests of one state or another should be given preference. No one of the present agencies can do this.

If these problems are not to lead to continued delays and litigation, some form of independent coordinating body must be created in the federal government. This coordinating agency would not have direct authority, but it could make appropriate recommendations to Congress, which would authorize the appropriate alternative, just as major river projects are authorized now.

Also, there is a different type of problem, for example, where the EPA is deciding upon the urgency of runoff control of pollution or TMDLs in a given watershed. In this case, the states (or the upstream state) may not agree with the degree of control required. This kind of problem has not yet arisen nationally because the EPA has not yet pushed its legal authority very far. However, the environmental interests may intervene to require the EPA to enforce the legal provisions literally, as they have in previous cases. That our federal administrators do not habitually grossly misuse available legislative power is not a justification for excessive powers to be granted, since it leaves an important field open for policy to be determined on an arbitrary basis, without adequate analysis.

Under such a situation, the EPA should conduct the study, but a national coordinating council should analyze the situation and report to

Congress, rather than leaving the problem to be decided by costly and prolonged lawsuits. This coordinating agency would perform much the same function as the Water Resources Council did while it existed. However, to avoid confusion, it would probably be best to give it a different name, perhaps Fed-State Water Council.

Consequences If No Major Action Taken

If Congress does not move to establish the system of cooperative Fed-state studies and the Fed-State Water Council as outlined, or some equivalent arrangement, the adverse consequences will develop over a period of time. If no further action is taken, as long as years of normal rainfall continue, the only noticeable change will be a gradual increase in the usage of water for water supply and other consumptive purposes, and a corresponding slow decrease in low flows downstream. Nonpoint-source pollution will also increase. Then one year a major drought will start. There will be a drastic reduction in flows downstream, widespread drying out of wetlands, and probably devastation of numerous endangered species. No extra storage or water conservation programs will be available to cope with such an emergency. Navigation, if any, will be seriously impaired. Cities will run short of water. Predictably, there will be an outpouring of public indignation, especially from environmental interests. Hopefully, however, it may be that remedial action will be taken before this situation arises.

References

Keyes, A.M. and Palmer, R.N. "An Assessment of Shared Vision Model Effectiveness in Water Resources Planning," *Proceedings of the 22nd Annual National Conference, Water Resources* Planning and Management Division of ASCE, Cambridge, Mass., May 1995, pp. 532-535.

Keyes, A.M. and Palmer, R.N. "The Role of Object-oriented Simulation Models in the Drought Preparedness Studies," *Proceedings of the 20th Annual National Conference,* Water Resources Planning and Management Division of ASCE, Seattle, Wash., April 1993, pp. 479-482.

Palmer, R.N., Keyes, A.M., and Fisher, S.M. "Empowering Stakeholders through Simulation in Water Resources Planning," *Proceedings of the 20th Annual National Conference,* Water Resources Planning and Management Division of ASCE, Seattle, Wash., April 1993, pp. 451-454.

Glossary

A

ACT-ACF study Alabama-Coosa-Tallapoosa, and Apalachicola-Chattahoochee-Flint study.

anadromous fish fish that live most of their lives in the ocean but ascend the rivers to spawn. The young fish then make their way back to the ocean.

aquifer porous layers underground from which water may be drawn.

B

benefit favorable effects of a given project or policy.

benefit/cost ratio (B/C) the economic valuation of tangible benefits due to a project compared to the economic valuation of its costs.

best management practices (BMP) the standard procedures and/or designs for a given situation.

biodiversity the main objective of environmental programs; the value of the diversity of living species.

C

C.S.O. combined sewer overflow. Combined sewers, in some older cities, carry rainfall runoff as well as sewage, and when they overflow during a storm, they release highly polluted water.

comprehensive water resource planning planning that considers not one but all purposes for which water is used or controlled. (Many "comprehensive" plans of the past failed to consider aspects now considered important.)

conjunctive use a water supply system combining water from surface sources with groundwater.

D

desalination processing of saline waters to produce fresh water.

detention dams or basins small impoundments built to reduce flooding in small streams, and, in many cases, also to reduce pollution in the runoff.

dynamic equilibrium characteristic of the environment by which natural changes, occurring over a period of time, are countered by natural resilience of the species concerned.

E

endangered species a variety of plant or animal that is becoming much more scarce and is officially protected by the government.

F

fed-state study a proposed federally authorized cooperative study.

floodplain management the control of land use in the floodplain so as to minimize future flood damages, and preserve environmental advantages.

I

intangible benefits benefits that cannot be evaluated accurately in monetary terms.

M

maximum contaminant levels (MCL)   levels of contaminant that must not be exceeded at the point in question.

N

Native Americans   aboriginal inhabitants of North or South America.

nonpoint-source pollution   the dispersed pollution that comes from streets, automobiles, homes and gardens, pets, and other small sources. Unfortunately, the Clean Water Act includes all of such pollution as point sources if it reaches the stream through a pipe, ditch, or channel, as most of it does. This latter definition is seldom used.

NPS   nonpoint-source pollution.

O

ozonation   water treatment process in which the greater part of the chlorine used is replaced by ozone

P

"petting zoo"   facility with animals that children may touch and per-haps feed.

pumped storage plant   hydroelectric power plant that operates in two modes. It can pump water up to a higher elevation, then recapture most of the energy involved by running the water back down through a tur-bine to the lower elevation.

R

riparian rights   rights to use water from streams where no system of water rights or water allocation has been instituted. They are based on court decisions of the remote past.

river basin commission coordination agencies established by federal authority but now operative in only a few river basins.

runoff pollution control the control of precipitation runoff both in order to reduce flood damage, and to settle out particulate pollutants. It is the old program of stormwater management, with control of pollutants added.

S

shared vision modeling the use of computer modeling in complex studies, where each participating agency provides a computer expert. These experts jointly design and implement the computer approach.

stormwater-treatment wetlands wetlands constructed to remove runoff pollution.

sustainable development development that does not include hidden factors that adversely affect the future.

T

technology-based programs programs that base their degree of control upon the characteristics of the site, rather than upon an area-wide analysis of the particular location.

total maximum daily loads (TMDLs) the total of a pollutant allowed in a given section of river or stream.

W

water rights legal system of entitlements to use water in the Western states.

water supply critical area a management device whereby the state enforces a proportional reduction in all groundwater withdrawals, thereby reducing total withdrawals to a level that can be maintained by natural recharge.

wellhead protection areas designated areas adjacent to a well, within which polluting activities are excluded.

wet basin detention basin designed so as to allow percolation into the ground, for recharge of groundwater.

wetlands lands characterized by vegetation that grows only where frequently flooded.

Index

A

Absolute and relative values, 23-24

ACT-ACF problem and study, 100-101

Anadromous fish, 66

Areawide planning, 82

Basic objectives of planning, 3-6

Benefits:

      environmental, 11-13, 15, 20-21, 70

      flood control, 57-58

      navigation, 54

      hydroelectric power, 67-68

      irrigation, 70-71

      water supply, 32

B

Basic policies of coordination, 12, 98

Benefit evaluation, general, 3-6, 98

Biodiversity, 12-14

Birth rates and population, 27-28

Bureau of Reclamation, 1, 8, 68-69

Drought management, 45-48

Dual purpose detention basins, 87-92

Dynamic equilibrium, 13-14

E

Ecosystem management areas, 21

Encroachment upon open space, 28

Endangered species, 13, 22-24

Environment, general, 1, 3-4

Environmental effects downstream, 21-22, 65-67

Environmental impact statement, 7

Environmental objectives as absolutes, 23-24

Environmental use of floodplain zoning, 60-61

Environmental Protection Agency (EPA), 6-8, 79-83, 99, 102, 107-108

EPA, new programs, 79-81, 92-93

F

Federal and state responsibilities, 1, 2, 6-10, 97

Federal requirements, runoff control, 92-93

Federal-state coordination (proposed), 105-109

Fish and dams, 5, 21-22, 65-67

Flood control, 51-61

Floodplain management, 53-56, 60-61

Floods, 51-61

Florida runoff control, 94

G

Gilbert White, 54

Glen Canyon Dam, 67

Government agencies (federal), 6-10

Government attempts at coordination, 97-103

Great Lakes, 6-7

Groundwater and conjunctive use, 34-35

H

Human welfare, 4-6

Hydroelectric power, 64-68

Hydroelectric power, controversies, 65-67

I

Industrial pollution control, 80, 81

Intangibles, 3-6, 77

Irrigation in the West, 68-69

Irrigation other than in the West, 69-71

Irrigation, 68-71

M

Manasquan dam, 17

Maryland runoff control, 93-94

Massachusetts Water Resources Authority, 46

Maximum contaminant levels, 81

Missouri River, 73-74

Missouri-Mississippi 1993 flood, 51

Money values for intangibles, 6, 23

Multiple purpose water resource planning, 73-77

Municipal runoff control, 92-93

N

National estuary studies, 99

National study of drought management, 47-48

Native Americans, 5

Navigation, 63-64

New Jersey runoff control, 94-95

New Jersey water supply, 36-37, 39, 40-45

Nonpoint-source pollution (runoff pollution), 29, 79-81, 85-96

NPS pollution and water supply, 88

NPS pollution coordination, 98-99

NPS pollution from agriculture, 87-88

NPS pollution removal methods, 87-89

Nutrients (nitrates) as a pollutant, 35, 86

O

Objectives of water resource development, 3-6, 98

Ogallala aquifer, 34

Ohio River, 74

Open space, loss of, 28

Overall coordination policies, 98

Ozonation, 35-38

P

Parks and forests, 12, 20-21

Petting zoo, 21

Pick-Sloan plan, 73-74

Platte River coordination, 102

Pollution and groundwater depletion from irrigation, 70

Pollution control and water quality, 79-83

Population and development, 27-29

Potholes reservoir, 20-21

Potomac River, 40

Pumped storage, 65

R

Rapid City, S.D., 13, 19

Regional detention basin systems, 15-16, 92

Relative and absolute values, 23-24

Reprocessing wastewater, 37

Restoration of wetlands, 16-18

River basin commissions, 9-10, 99, 101

Rivers, lakes and reservoirs, 32-33

Runoff pollution control, 85-96, 108

S

Safe Drinking Water Act, 32, 81

Salmon and dams, 5, 65-67

Septic tanks, 34-35

Shared vision modeling, 105-107

Special districts (flood control), 57

State programs, runoff pollution control, 93

State responsibilities, 97, 99, 107-109

Storm-water management, 85-96

Subsidies to irrigation, 68-69

Sustainable development, 5

T

Tax-equivalent basis, hydropower valuation, 67-68

Tennessee Valley Authority (TVA), 73

Tocks Island Dam, 14

Total maximum daily loads (TMDL), 7, 93

Tri-County Water Supply Project, 35-38

U

U.S. Army Corps of Engineers (see *"Corps of Engineers"* **),** 1

Urban Drainage and Flood Control District, 57

Urbanization, 28

V

Valuation of benefits (see *"Benefits"*)

Values, relative and absolute, 23-24

W

Water conservation, 46-47

Water pollution control, coordination, 82-83

Water quality and pollution control, 21, 79-83

Water quality criteria and standards, 81-82

Water resources
    history, 1
    objectives, 3-6
    future trends, 29

Water Resources Council, 10

Water supply critical areas, 40-45

Water supply interconnections, 39, 40

Water supply, 2, 31-48

Water treatment, 35-37

Water uses and effects, 2

Wetlands control, 8, 15-19

Wisconsin runoff control, 94

World birthrates and immigration, 27

Y

Yolo Bypass State Wildlife Area, 18